Roses
without chemicals

150 disease-free varieties
that will change the way you grow roses

Roses
without chemicals

Peter E.
Kukielski

Timber Press

Portland London

Frontispiece: 'Alexandra Princesse de Luxembourg'

Published in 2015 by Timber Press, Inc.

Illustrations on pages 34-35 by Peter Kukielski, produced by David Jacobson.
Photo credits appear on page 258.

The Haseltine Building   6a Lonsdale Road
133 S.W. Second Avenue, Suite 450   London NW6 6RD
Portland, OR 97204-3527   timberpress.co.uk
timberpress.com

Printed in China
Text and cover design by David Jacobson, ORT

Library of Congress Cataloging-in-Publication Data

Kukielski, Peter.
Roses without chemicals: 150 disease-free varieties that will change the way you grow
roses/Peter E. Kukielski.—1st edition.
     pages cm
Other title: One hundred fifty disease free varieties that will change the way you grow roses
 Includes index.
 ISBN 978-1-60469-354-6
1. Roses—Varieties—North America. 2. Roses—Disease and pest resistance—North America.
I. Title. II. Title: One hundred fifty disease free varieties that will change the way you grow roses.
SB411.6.K85 2015
635.9'33734—dc23   2014020741

A catalog record for this book is also available from the British Library.

For Drew

Contents

Growing roses sustainably 62

The chemical-free rose directory: 88
150 disease-resistant roses

Roses by class, habit, and color 242

Preface
(It's not your fault)

Beautiful, healthy roses like 'PlumPerfect' are part of a new trend toward sustainable rose gardening.

Whether you are a home gardener or the steward of a public rose garden anywhere in the world, I want you to have the confidence to grow roses, or to grow roses again, without chemicals. That's my dream and that's why I wrote this book. By the time you have finished reading, I hope you will feel free to grow a huge variety of these spectacular plants.

Because nearly everyone has heard the phrase "Rose is a rose is a rose is a rose" I often quote it when talking with people in the Peggy Rockefeller Rose Garden at the New York Botanical Garden, where I was the curator for eight growing seasons. These words come from the 1913 Gertrude Stein poem *Sacred Emily,* and people have taken the line to mean something like "Things are what they are." Ironically, the rose was a very bad example for Stein to use for her metaphor. Taken as commonly understood, the sentence would mean that all roses are basically the same, and no matter which pretty picture you see in a rose catalog, the plants are all going to grow the same, smell the same, and perform the same. Stein would have been more on target if she had written, "Rose is (not) a rose is (not) a rose is (not) a rose." That's because all roses are not created equal. Or, more importantly, all roses are not created for the same purpose.

The process of creating new rose varieties is called hybridization. Breeders cross one rose with another rose to create a new variety that has a different combination of genes than either of its parent plants. Almost all roses that you can buy today have been hybridized for one purpose or another. Sometimes that purpose is to emphasize a gorgeous color that catches your eye. Maybe the hybridizer likes a certain cupped flower. Sometimes it is for a particular fragrance or growth habit. Some roses are

created with hardiness in mind, because the hybridizer wants or needs the roses to survive harsh winters.

With thousands of roses now available on the market, the choice to the home gardener can be daunting and confusing. In my quarter-century of purchasing and growing roses, I have always desired to find a rose that is better than the one I am growing at the moment. I'm always on the lookout for the next best thing, the next best rose. Does this sound familiar? Some of my friends have a similar desire with fashion—always wanting the next great trend or popular item. I used to think that roses are similarly fashionable and that the rose industry mirrored the fashion industry. In both these worlds, a color or style can be hot one year and out the next.

Yet too often, when I found a stunning image of a rose in a magazine and determined that I must have that treasure, ordered it, put it in the ground, cultivated it, and loved it—it rewarded me with disappointment. The leaves became diseased, its color or fragrance was lackluster, or even worse, the entire rose bush died. Many despondent and frustrated rose lovers have shared similar stories with me.

Perhaps this has also happened to you and if so here is the central point I want to make in this book: it is not your fault. In the pages that follow I am going to explain to you why some roses fail to thrive, and how to choose

and grow roses in an environmentally sensitive way for your garden, in your part of the country. In the directory you'll find 150 of the best-performing and most disease-resistant roses available on the market today. I have grown every one of these roses myself and have chosen them out of the many thousands of other roses that I have grown and trialed over the years. I have included a rating for each rose based on the qualities that matter most to gardeners: disease-resistance, bloom, and fragrance. You can rest assured that they are the very best choices for a sustainable, chemical-free rose garden.

'Autumn Damask'

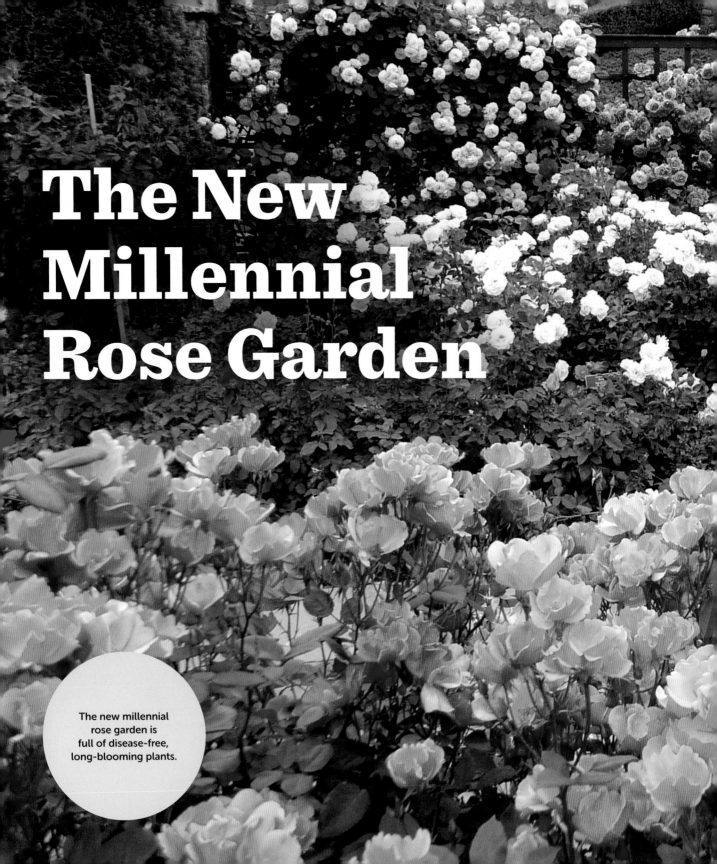

The New Millennial Rose Garden

The new millennial rose garden is full of disease-free, long-blooming plants.

A rose is hybridized for whatever purpose or purposes its creator is seeking, those qualities the hybridizer wants to maximize. But when a rose is hybridized to maximize any one quality, there is the possibility that some other facet will be compromised or sacrificed. Too often in today's marketplace, roses are hybridized for a narrow, superficial beauty that will attract the consumer in a catalog, garden center, or florist shop. But just like the fruit and vegetables that are bred to look perfect on supermarket shelves, these hybridized plants can go bad very quickly. Selecting roses because they have good looks may actually be counterproductive. That lovely rose may soon be riddled with leaf spot because the ability to resist disease has been bred out of it.

This book will help you learn about the specific hybridization efforts toward disease resistance and sustainability in roses. Of the thousands of roses available on the market, I want you to know about roses that are right for you and that you will be able to grow successfully, disease-free and chemical-free. I also want you to know about the best way to plant and care for these roses.

Right for you also means right for the area where you garden. Understanding the effect of your local climate on roses determines how successful you will be in growing them. It's unreasonable to assume that roses that might be successful in Miami or England would also be successful in Maine or Norway. If together we can identify roses that are good performers for your region and climate, then I know you will have better, healthier roses based on that factor alone.

I like to use the term "new millennial roses" for those varieties that are fragrant, beautiful, disease-free, chemical-free, relatively maintenance-free, and regionally appropriate. I believe that sustainable rose gardens are available to every home gardener, and to large commercial growers and public gardens everywhere.

I chose the word *millennial* to reference the year 2000. It was around this time that we—growers, breeders, and curators—began to shift our way of thinking about roses. We started to move in the direction of more "green" and sustainable roses. We began to look for roses that were hybridized specifically to be resistant to diseases. These are the roses that you will find in this book.

The Peggy Rockefeller Rose Garden

Just as the millennial shift in roses was happening, I became curator of the Peggy Rockefeller Rose Garden, which is located in the New York Botanical Garden. First laid out in 1916 by the eminent American landscape architect Beatrix Jones Farrand and nestled among beautiful, established trees, the site offers some of the most breathtaking vistas available at the botanical garden. Thanks to a generous gift from David Rockefeller in honor of his wife, Peggy, the garden was completed and named for her in 1988. With continuing support from Mr. Rockefeller, I had the honor of renovating the garden through the winter of 2006 and 2007. The garden was reopened in 2007, and when I left in 2013 it contained a significant living display of more than 4000 roses and close to 700 different varieties.

The rose garden covers just over an acre and is triangular in shape, with a circular central area containing a focal gazebo. When I took over, the rose garden had been sprayed with chemicals for twenty years. The collection had about 2000 roses in it, approximately 234 varieties. Originally, the goal of the renovation was to increase diversity and make better use of the planting space. During the initial renovation, I took out 400 roses and added 1700 new roses, almost doubling the size of the collection.

As part of the renovation the beds were redefined in order to diversify and reorganize the rose classes in the existing collection.

The rose garden at the New York Botanical Garden has impressed visitors since its creation a century ago.

On the southeast side, we created a heritage rose border that features a chronological sampling of the history of roses, showing their lineage as they were hybridized over the centuries. It begins with species roses and their cultivars, wild plants that evolved through natural selection, and continues down the timeline of rose history from Gallicas and Damasks to Albas, Centifolias, and China roses, from Spinosissima shrubs to Moss roses, Portlands, Bourbons, and finally Hybrid Perpetuals. Two other borders are planted with modern roses. Signage tells the story of the development of the rose, from the antique plants to the more recent varieties.

After the initial renovation, 2008 presented the rose garden with a new mission. I wanted to modify the collection to even better represent the diversity of roses and also to display great garden plants for the public to learn about and enjoy. As new plants were being hybridized, trialed, and introduced from around the world, I felt we should bring to the forefront those that are highly disease-resistant and easy to grow. At the same time, we wanted to preserve important historical rose varieties of the past for educational purposes.

As we went through the growing seasons from 2009 to 2013, we looked for ways to reflect these new millennial roses. With this in mind, we planted revised modern collections that included more disease-resistant varieties such as those from the Texas Pioneer, Griffith Buck, and Easy Elegance series, as well as those with the

Signs throughout the garden tell the history of roses from their wild ancestors to the present day.

A rose named Eva

In June 2007, after the extensive renovation of the Peggy Rockefeller Rose Garden was completed, we reopened the garden. The spectacular first bloom was under way. I was in the garden talking with some visitors when a young woman with an abrupt manner approached me.

"Good afternoon!" I greeted her.

"Where is Eva?" she asked impatiently, her foot tapping on the ground.

I was a little shocked at the stern expression on her face. But then I guessed why she was perturbed.

"I bet your name is Eva," I replied.

"Yes," she said, still waiting for an answer. "Where is my rose?"

Fortunately, I was able to direct her to the area of the garden where we grow Hybrid Musks. She walked briskly over to the bed and was able to see that Rosa 'Eva' was blooming beautifully, with endless clusters of blended reds.

Once satisfied, Eva (the human) proceeded back in my direction as quickly as she had left and said, "Thank you. She looks good! I'll be back again next year—make sure you take good care of her!"

"Oh, I will!" I said, almost fearing for my life while I pledged to protect her personal floral symbol.

Having received direct orders from this dynamic young woman, it became clear to me at that moment that this was not only my garden to care for but it was everyone else's garden as well. I watched visitor after visitor come into the garden and stake their special claim to a favorite rose or two (or three or more), and derive immense joy from this public space that they saw as their own. 'Eva' and the other nearly 700 varieties must be well taken care of for the enjoyment of one and all.

Earth-Kind designation. We also put in extensive plantings of roses that emerged from the hybridization efforts from Kordes of Germany, Meilland of France, and Will Radler of the United States, along with several of the winners from the ADR Rose Trials in Germany, which include some of the best hybridization efforts in the world.

In the last eighty years, there has been a movement away from the use of roses as general garden plants. In part, this can be attributed to the level of culture required to care for roses that were bred without regard for the health of the plants. But things are changing with the new millennial roses. Consumers are willing to pay premium prices for cultivars that are resistant to common diseases, easy to maintain, and ever blooming. I can say with great enthusiasm that modern hybridizing efforts are really putting the rose back to its rightful place as a great garden plant.

In June 2010, the Great Rosarians of the World (GROW) presented the Peggy Rockefeller Rose Garden with the International Rose Garden Hall of Fame Award. The same year, the All American Rose Selections (AARS) declared the garden one of the best public rose garden displays in America, conferring this honor "in recognition of the Creation of a Sustainable Public Garden Representing an Outstanding Collection of Historic Roses." In the fall of 2012, The World Federation of Rose Societies presented the Garden with the prestigious Award of Excellence recognizing it as one of the best rose gardens in the world.

The garden is under the watchful eye and care of a team of horticultural professionals with many years of very specific rose experience. It is with proactive, consistent care—the use, for instance, of compost, mulch, organic fertilizer, and sustainable growing methods that I will be discussing in

Many of the roses in the Peggy Rockefeller garden have demonstrated their ability to resist disease.

this book—that the health of the garden can be kept at its highest.

I remember walking through the rose garden with David Rockefeller on one of his visits (he used to visit each year on his birthday). At one particular moment, this kind and gentle man turned to me and whispered, "Peggy would have loved this." Although this gave me a sense that our mission had been accomplished, I also realized that the garden must continue to evolve its mission. As with all gardens, its caretakers must constantly look for new ways to make this garden as beautiful and sustainable as it can possibly be, so that visitors have the same experience of loving it year after year.

The Evolution of Roses

I believe that you can't have a book about sustainable roses without mentioning those roses that have grown, survived, and proven themselves over time without chemicals. Many of these are heritage roses, also called antique or old garden roses, and I urge you to look further to explore the great diversity among them. Here I want to give you a general sense of how the world of roses has expanded over the last few hundred years. This includes how and why roses became the chemically dependent plants that the modern gardener has known them to be over the past few decades. In order to understand this better, it is helpful to look at the development of roses over the centuries.

Roses have grown wild on this earth for millions of years. A fossil of a rose has been found that dates back 34 million years. And for 34 million years, these plants were not treated with any fungicides that I am aware of. Just think about it: these tough plants have been growing throughout the Northern Hemisphere, thriving on no maintenance and no intervention from humans. The history of the cultivated rose includes records from the Egyptians to the Greeks and Romans, all the way to Europe in the 1700s, when growers started to develop official classes of roses. Any rose that was hybridized before the year 1867 is considered a heritage rose. Any rose after that date is considered a modern rose. I have coined the phrase "new millennial" for those varieties post-2000 that are specifically hybridized for or have a proven propensity toward disease resistance.

SPECIES ROSES

The species roses (with some exceptions) have a "simple" flower form. The bloom opens flat and has only five petals. The family Rosaceae, which includes the species roses, also includes other plants such as apples, cherries, and pears, as well as strawberries and even ornamental shrubs like Kerria and Spiraea. If you look at the blossoms of these species, they all have the simple flower form of five petals.

Some of my favorite species roses include

Rosa blanda

Rosa canina

Rosa chinensis

Rosa gallica

Rosa hugonis

Rosa nutkana

Rosa roxburghii

Rosa sericea

Rosa spinosissima

Rosa virginiana

Species roses bloom only once per year, covering themselves with flowers in the late spring and early summer. The flowers are followed by colorful rose hips that provide food for birds and winter interest in the land-scape. Estimates vary, but there are probably up to 150 different rose species.

If nature provided this many species roses, what happens when one rose is crossed with another? Simply speaking, a hybrid cross is made—a rose with different genes than both of its parents. This cross may occur naturally when bees or other polli-nating insects disperse pollen from bloom to bloom, or they can be man-made by purpose-fully taking pollen from the stamens of one plant and applying it to the stigma of another plant to create new seed, then growing plants from that seed.

Any existing rose can be crossed with any other existing rose to come up with a new hybrid rose. If you cross a five-petaled flower with another five-petaled form, you might get five-petaled offspring or you might get offspring that have flowers with more petals. Now think about what might result if you crossed this multi-petaled rose with

Rosa chinensis

Rosa canina

Rosa gallica 'Officinalis'

Rosa virginiana

'Cardinal de Richelieu'

a five-petaled rose. It might possibly be a new form of many-petaled flowers that does not look like the original species rose. This is how a new class of roses is born. A broad definition of a class of roses is that they share a common flower form. A class is considered distinct because the blooms are different from those of the parent plants.

HERITAGE ROSES

Heritage, antique, or old garden roses are the first roses that were brought into cultivation. Their history is a journey through the centuries, as gardeners and breeders sought to select and then to hybridize new and exotic types of rose.

Gallica | *Rosa gallica* is a species rose that is native to southern and central Europe. The oldest named form of this plant is *R. gallica* 'Officinalis' (also known as the apothecary's rose, because it was thought to have medicinal properties) and it dates from as early as the 14th century. *Rosa gallica* 'Officinalis' has multiple petals, as do other descendants of *R. gallica,* sometimes up to 100 petals per flower. This class of roses became known as the *Gallicas.*

Rosa gallica 'Versicolor'

Damask | The next new class of roses were the Damask roses. The original Damask rose was thought to be a cross between Rosa gallica and another species of rose, but many experts argue over this point. The Damask roses do not have as many petals as the Gallicas before them, yet they have many more petals than the earlier species. The Damasks are most known for their fragrance and are a main source for rose oil or "attar of roses" for the perfume industry.

Some of my favorite Damask roses include

'Autumn Damask'

'Ispahan'

'Kazanlik'

'Leda'

'Madame Hardy'

Alba | Next to be developed was the Alba class of roses. Alba means white, but the

'Leda'

'Ispahan'

Albas actually include roses that range from white to pinks. They have a wider variety of flower forms and a lighter fragrance than the heavier Damask scent of their predecessor. Another characteristic of this class is that the foliage shares a glaucous (grayish-green or blue) coloring that differs from the brighter green foliage of earlier classes.

'Alba Semiplena'

Early Alba roses are thought to be descendants of the species *Rosa canina* crossed with *R. damascena*.

Some of my favorite Alba roses include

'Alba Semiplena'

'Félicité Parmentier'

'Great Maiden's Blush'

'Königin von Dänemark'

'Madame Plantier'

'Pompon Blanc Parfait'

'Sappho'

'Félicité Parmentier'

Centifolia | A simple qualifying characteristic of Centifolia roses can be gleaned from the name *centi* meaning

'Dometil Beccard'

hundred. These blooms, which are quite different from other classes of roses, are generally very cupped or rounded in shape and do indeed have about 100 petals. Centifolia roses are said to have been the first class developed as late as the 17th century. Maybe because of this early date of cultivation, I have found the number of Centifolia roses in commerce to be fewer than other classes. The ones I have grown have been a wonderful addition to the garden.

Some of my favorite Centifolia roses include

'Cristata'

'Dometil Beccard'

'Fantin-Latour'

'Le Rire Niais'

'Rose de Meaux'

'Fantin-Latour'

Moss | Moss roses have bloom characteristics similar to those of their predecessors, and so they can be difficult to identify simply by looking at the flower form. The obvious difference is the addition of "mossy" glands to the buds, which changed this flower form and so distinguished the class. The flower buds are literally covered with glandular tips that produce their own oily scent, which can range anywhere from citrus and anise to earthy notes, which make a wonderful complement and contrast to the rose scent of the flowers. The Moss roses are known to be sport descendants of the Centifolias.

'Comtesse de Murinais'

'Gloire des Mousseux'

Some of my favorite Moss roses include

'Capitaine John Ingram'

'Comtesse de Murinais'

'General Kleber'

'Gloire des Mousseux'

'Jean Bodin'

'La Diaphane'

'Madame Louis Lévêque'

'Salet'

'William Lobb'

'La Diaphane'

China | Although there is much to say about China roses, the simple fact is that they come from China and they were developed from the species *Rosa chinensis*. China roses are among the most floriferous of all roses and are credited with having given most modern roses their remontant qualities. They also brought new flower forms and colors to cross with existing European roses. The China roses include shades from white to pink, and also brought some pastel shades and some darker reds, crimsons, and even purples to the world of roses. Any rose called a hybrid China would be a cross between the China roses and other existing classes of roses. I have included two China roses ('Ducher' and 'Mutabilis') in the directory because, like several other heritage roses included there, they have proven their disease resistance and blooming capacity in gardens for many years.

'Archduc Charles'

Some of my favorite China roses include

'Archduc Charles'

'Arethusa'

'Cramoisi Supérieur'

'Ducher'

'Mutabilis'

'Old Blush'

'Viridiflora' (the green rose)

'Old Blush'

'Viridiflora' (the green rose) with 'Gourmet Popcorn' (white)

'Comte de Chambord'

'Indigo'

'Rose de Rescht'

What is a sport?

A sport is a genetic mutation that occurs innately and leads to a difference in part of a plant. The mutation is often a color sport, such as the appearance of a pink flower on a white-flowered rose. A growth habit sport may occur when a rose has a compact bush growth habit but then one of the branches elongates and develops as a climbing form. A blooming sport can occur when a single-blooming rose starts to repeatedly flower throughout the year, which is referred to as remontant. To retain that new characteristic the changed element must be selected from the plant and propagated to create new plants with those characteristics. In the example of Moss roses, the sport went from a simple flower bud to one that had glandular moss covering the bud.

Portland | Because roses inherit the genetic attributes of all their ancestors, the number of genetic and possible hybrid crosses available throughout the world of roses grows exponentially. Sometimes it can be difficult to tell certain classes apart, because they share so many heritable characteristics. Portland roses are a case in point, and I find it difficult to distinguish this class of roses from classes that preceded them. Portland roses offer the repeat blooming characteristic that traces back to their China rose heritage. They also offer some of the same classic flower forms as their previous parent classes of Gallicas and Damasks. One unique characteristic of Portland roses is that their bloom is "presented" with a grouping of leaves just below the flower, much like a nosegay. This effect can be detected easily in some varieties but less so in others.

'Edith de Murat'

'Variegata di Bologna'

Some of my favorite Portland roses include

'Comte de Chambord'

'Indigo'

'Marchesa Boccella'

'Rose de Rescht'

'Rose du Roi'

Some of my favorite Bourbon roses include

'Commandant Beaurepaire'

'Edith de Murat'

'Eugène Desgaches'

'Honorine de Brabant'

'Louise Odier'

'Madame Ernest Calvat'

'Madame Isaac Pereire'

'Souvenir de la Malmaison'

'Variegata di Bologna'

'Zéphirine Drouhin'

Bourbon | With roots that link back to the China roses and the Damask roses, Bourbons have the qualities of both. From their China ancestors, the Bourbons seem to have gained a broad range of colors, a delicacy in their blooms, and good reblooming ability. From the Damasks, the Bourbons show some flower form characteristics and deep fragrance. Bourbons are very easy to find in commerce and are wonderful for the scented garden. Although I have seen some black spot on Bourbons in the extreme humidity of the summers in some areas of the country, this class of roses generally has all the characteristics that the home gardener desires.

Noisette | The Noisettes are some of the great flower producers in the rose garden. The genes for powerhouse blooms come from their China parentage, and their wonderful fragrance and clustering of flowering comes from the species *Rosa moschata*. The origins of this class date back to 1802 when John Champneys of South Carolina developed some seedlings and Philippe Noisette further developed them in France. Although the Noisettes are grouped in with these other heritage roses, I have included a few of them in the directory ('Alister Stella Gray' and 'Blush Noisette') because they have consistently proven their disease resistance and blooming capacity over and over again in the garden.

'Blush Noisette'

Some of my favorite Noisette roses include

'Alister Stella Gray'

'Blush Noisette'

'Champneys' Pink Cluster'

'Duchesse de Grammont'

'Fellenberg'

'Madame Plantier'

Hybrid Perpetual | Every time I teach a class on heritage roses, by the time I get to the Hybrid Perpetuals, the chalkboard and overhead projector combined look like a jumbled mess of words, descriptions, pictures, and sketches. Like my scribbled chalkboard, the Hybrid Perpetual roses are a combination of all the rose genetics before them. Within the complexity of this class, a new flower form emerged that tended to be rather large, full, and higher-centered than previous classes. This voluptuous bloom got the attention of breeders and customers alike, and people began to make crosses of roses more than ever before. However, due to the jumble of genes that have been bred into cultivated roses by this point, and the strict focus on perpetuating above all else the genes that determine flower form, this is also the first class of roses that began to lose the heritable disease resistance of its ancestors.

'Madame Plantier'

MODERN ROSES

'La France' was hybridized in 1867 by Jean-Baptiste Guillot (1803–1882) by crossing a Hybrid Perpetual and a Tea rose (tea-scented China rose, *Rosa ×odorata*). 'La France' is generally accepted to be the first Hybrid Tea rose in Europe, making this date the dividing line between heritage and modern roses.

Hybrid Tea | Hybrid Teas are arguably the most numerous of roses and many new varieties enter the market each year. They are remontant, free-branching shrub roses of upright or bushy habit, with prickly stems and glossy or matte, mid- to dark-green leaves. They have one large, usually double, sometimes scented bloom per stem. The desire for these flowers even spurred a new industry—production and selling of the florists' rose. Have you ever been given a dozen roses on Valentines Day? They are undeniably beautiful, but if you sniff the bouquet, what is the result? Although the first thing most people do with roses is smell them, with these flowers there is likely to be little fragrance.

You will notice there are very few Hybrid Tea roses in the directory. That's because, like Hybrid Perpetuals, many Hybrid Teas have lost the natural disease-resistant genes of their more sturdy ancestors. Bred for a single characteristic—a specific type of flower form—most of them have lost many of the other characteristics that make a rose suitable for a sustainable rose garden. Fortunately, with the renewed interest in breeding sustainable roses, we are starting to see the introduction of more fragrant, disease-resistant Hybrid Teas.

Polyantha | Typically compact shrub roses with prickly stems and glossy green leaves, Polyanthas are usually remontant. Sprays of

'Louis van Houtte'

'Marchesa Boccella'

'Peace' is one of the most famous Hybrid Tea roses in the world.

'The Fairy', like most Polyanthas, produces flushes of small double flowers.

'Mother of Pearl'

small, single to double flowers are produced in flushes from late spring to autumn.

Floribunda | This class resulted from crossing a Hybrid Tea with a Polyantha. Floribundas are free blooming, free branching shrub roses of upright or bushy habit, usually with prickly stems, and glossy, green leaves. Single to fully double, sometimes scented flowers are usually in clusters of three to twenty five, and are produced continuously from late spring to fall frost. A climbing sport of a Floribunda would be a Climbing Floribunda.

Grandiflora | This class results of a cross between a Hybrid Tea and a Floribunda. Grandifloras are remontant, free-branching shrub roses of upright or bushy habit, with prickly stems, and glossy or matte, mid-to dark-green leaves. Large, usually double, often scented flowers are generally solitary although sometimes produced in clusters. The first Grandiflora was 'Queen Elizabeth'.

'Betty Prior' and 'Sans Souci' display the clusters of multiple blooms that characterize the Floribunda class.

"Roses are difficult!"

The two most common phrases I hear from home gardeners are "I don't want to spray" or "I don't want to use chemicals." Unfortunately what often comes right after these phrases is "That's why I don't grow roses. They're too difficult."

Like many gardeners and rose growers, I am aware that sometimes the right chemical at the right time can help to solve problems in the garden. But the fact is that many home gardeners have been purchasing roses for their gardens that were doomed to fail because the plants were genetically unsuitable for the growing conditions in their region. When you think about it, why are the same roses sold in Florida, Colorado, and Oregon?

The conversation often continues, "Look, Peter, I just want to go to the garden center or on the internet and trust that the roses I am buying will do well in my backyard."

I often say that the best way to learn about roses is to grow roses. It can be as simple as that. By trial and error, you could find out what roses grow well in your garden. But with the multitude of choices in the garden center shelves, it is difficult even to know which ones

to trial. That's why the industry started to breed, grow, and sell roses that do well regionally. By now, a lot of this trial-and-error has already taken place.

Fortunately, if you look at the garden center shelves these days you are likely to see roses that have been bred to be reliable for home gardeners. Gardeners are making the shift to low-maintenance roses and more environmentally responsible landscapes. This is illustrated by the success of the Knock Out series of roses developed by American amateur hybridizer Will Radler (the original cultivar in the series was introduced in 2000). The level of disease resistance coming from the Kordes and Meilland hybridization efforts also continue to impress me and they consistently score at the highest levels within the Peggy Rockefeller Rose Garden. To date, 'Knock Out' and its siblings are among the best-selling plants in rose history. I can't tell you how many people have told me they grow 'Knock Out' in their garden with success. What a great turnaround from telling me that they "can't grow roses" or that "roses are too difficult!"

The first of the Grandiflora roses to be hybridized was 'Queen Elizabeth', introduced in 1954.

Large-flowered climber | These are vigorous climbing roses with arching, stiff canes that are typically covered with prickly thorns, and often dense, glossy green foliage. Many have scented flowers in a variety of forms, borne singly or in clusters. Some bloom in spring or early summer only, on short shoots that emerge from the previous year's canes; others are remontant and flower on new canes.

Groundcover | These are spreading and trailing shrub roses, mostly with prickly stems and glossy leaves. They bear clusters of numerous single to fully double, sometimes scented flowers. Some flower in summer only, and many varieties produce flowers all along the stems.

Hybrid Musk | Hybrid Musk roses are vigorous, spreading shrubs that bear graceful blooms in clusters. The blooms are highly scented and repeat all season long. These roses are best when grown freely to let their natural shape develop in the landscape. Many roses in this class have a high degree of disease resistance.

Rugosa roses are known for the beautiful hips they produce in autumn.

Hybrid Rugosa | Shrub roses with tough, wrinkled, and usually bright green leaves, Hybrid Rugosas are strong, hardy roses. Most bear clusters of single or semi-double, scented flowers. Rugosas are wonderful at producing large, colorful rose hips. Although many do not typically rebloom very well, they are generally disease-resistant.

Miniature | These are compact shrub roses with very short stems and small flowers and leaves. The blooms are remontant and may be produced singly or in small clusters. These are often good subjects for containers.

Shrub | This is a diverse, catchall group of shrubs that are usually larger than Hybrid Teas, with prickly stems and medium-size leaves. The flowers are often scented, single to fully double, and produced in few to many flowered clusters from late spring to autumn.

Some well-known series of shrub roses that you will find in the directory include

Carefree Roses

David Austin English Roses

Easy Elegance Roses

Griffith Buck Roses

Knock Out Roses

Kordesii Roses

Oso Easy Roses

Thrive! Roses

Arching

Bushy

Rose growth habits

Class, fragrance, color, and flower form are not the only ways in which roses differ from one another, and so not the only reason you would prefer one rose to another. The growth habits of roses vary greatly, making them suited to different roles in the garden, from hedges to specimen plants, container subjects, or clambering up a trellis.

Arching | Simply stated, these roses send out canes that are long enough to arch over, filling the garden with soft, elegant lines. Blooms may emerge along the cane or at the ends, depending on the variety.

Bushy | A bushy, or shrubby, growth habit is consistent, tight, and uniform. I often think that roses like this would be a beautiful addition to the landscape even if they didn't bloom. Generally, I would describe these plants as uniformly rounded and full.

Climbing | The canes of these roses extend or elongate so that they are suitable for growing against structures like fences, arbors, and pillars. Climbers may be as little as 6 feet tall or extend up to 20 or 30 feet. Rambling roses are similar to climbing roses, with very elongated canes that tend to be skinnier and more flexible than those of the climbers. Ramblers are mostly once-blooming plants.

Spreading | Roses with spreading habits take up more horizontal space because they grow wider rather than taller. For example, a spreading rose might be 4 feet wide but only 2 feet tall.

Upright | These roses simply want to keep pushing their growth upward. They are tall plants that do not arch or branch out widely.

Climbing

Spreading

Upright

Roses in the garden

Public rose gardens typically have something that the home gardener doesn't have the luxury of—space. More often than not a sea of roses (sometimes thousands of them at a time) is laid out before the visitor. In a typical front or back yard you are unlikely to have space for thousands of roses. However, even in the smallest space, you have many options for designing your garden to display your roses to best advantage.

I hate to think of a garden with just a single rose! Grouping roses together can be as simple as combining varieties with like colors, or alternatively with complementary colors. I enjoy combining roses of different colors and flower forms, because the whole effect becomes greater than the sum of the individual parts. Grouping by complementary growth habits is also a good way to approach a rose planting. You can put

In a large garden, you can combine many different varieties of roses to create a sea of blooms.

groundcover roses near the front of a border, followed by some varieties with lower growth or spreading habit, followed by yet another variety that is more upright, and the whole bed can backed by some climbing forms.

Another option is to group roses by class, such as creating a cutting garden of all long-stem Hybrid Teas or Grandifloras and Floribundas. Keep in mind, however, that there can be many different growth habits within a class, so just planting a collection of a particular class can give you all sorts of shapes and sizes that end up looking like a bit of a mishmash.

COMPANION PLANTING

Combining roses with annuals, grasses, perennials, shrubs, and vines is a great way to create color combinations, to make more interesting and creative borders, and to attract beneficial insects into the garden. Roses have one of the longest seasons of bloom in the garden. Unlike tulips, for example, remontant roses can be in flower from late spring all the way to the first frosts—and sometimes even beyond. This means you have a long season to play with different color and texture combinations in mixed plantings, creating stunning effects of changing color and texture as the season progresses.

Plants for the mixed border should all tolerate approximately the same growing conditions, although there are often spots in a border that are sunnier or shadier than others. Extremely drought-tolerant plants, like many succulents, may find life in the border with roses a little to damp for their liking. You can design an automatic system that delivers more water to the area around the roses than to other nearby plants, but it's simpler to choose plants that like the same general environment. Don't choose spreading plants, or those with extra-vigorous growth, because they will try to crowd the roses out of the picture. Generally, don't plant anything closer than 1 foot from the base of a rose bush. Some flowering plants are especially valuable for attracting predatory insects into the garden, which is a real asset in a chemical-free environment. The predators may be spiders, beneficial mites, parasitic wasps, or aphid-eating beetles, all of which like to munch on or lay their eggs in some of the more pernicious rose pests. By choosing companion plants that are both beautiful and beneficial, you'll enhance the appearance and health of your roses.

Think about the overall shape and texture of any companion plants. Showy, dramatic plants like the taller euphorbia, delphiniums, and foxglove should be used sparingly so they don't visually overwhelm the roses. Perennials and grasses with airy textures such as *Verbena bonariensis,* baby's breath (*Gypsophila*), and Mexican feather grass (*Stipa tenniussima*) form clouds of color that softly mingle with rose blooms. Bright, daisylike flowers like asters, tickseed, and coneflowers add punch to hot-colored borders with similar strongly colored roses. Medium- and low-growing perennials are excellent for underplanting roses. These plants will also work with the mulch cover to cool the soil and help keep it moist. In most

Consider pairing roses with very different flower forms to create contrast, such as this 'Golden Fairy Tale' with white 'Escimo'.

Rosa 'Mutabilis', with its kaleidoscope of colored petals, adds highlights to a mixed border with pink poppies and the giant feather grass, *Stipa gigantea*.

Fifty great rose companions

Botanical name	Common name	Colors	Height	Comments
Perennials & bulbs				
Achillea	yarrow	orange, pink, red, yellow	medium	summer through autumn bloom
Agastache	giant hyssop	blue, orange, pink, purple, red	medium	summer through autumn bloom
Alchemilla mollis	lady's mantle	lime, yellow	low	early summer through autumn bloom
Allium		pink, purple, white	medium to tall	stroking, globe-shaped flowers; bloom spring to midsummer
Anemone	windflower	pink, white, yellow,	low to tall	summer bloom
Aster	Michelmas daisy	blue, pink, white, yellow	medium to tall	late summer through autumn bloom
Astilbe		cream, pink, purple, red, white	medium	summer bloom
Astrantia		green, maroon, pink, purple, white	medium	summer bloom
Campanula	bellflower	blue, mauve, pink, purple, white	low to medium	herbaceous and evergreen species; summer bloom
Coreopsis	tickseed	pink, red, yellow	medium	daisylike flowers; summer through autumn bloom
Corydalis		blue, pink, red, yellow	low to medium	spring and summer bloom
Dianthus	pink	orange, pink, red, white	low to medium	evergreen, often blue or silver foliage; summer bloom
Echinacea	coneflower	orange, pink, purple, white	medium to tall	bold daisylike flowers up to 6 inches across; summer and autumn bloom

Botanical name	Common name	Colors	Height	Comments
Euphorbia	spurge	chartreuse, green, lime, maroon, yellow, white	medium	unusual sprays of colorful bracts; spring through autumn bloom
Gaillardia	blanket flower	orange, red, reddish-brown, yellow	medium	showy, daisylike flowers; summer through autumn bloom
Gaura		pink, red, white	medium	light, airy texture; summer through autumn bloom
Geranium	cranesbill	blue, mauve, pink, purple, white	low to medium	tough, spreading plants; spring through autumn bloom
Geum		apricot, pink, orange, red, yellow	medium	bright flowers; spring through autumn bloom
Gypsophila	baby's breath	pink, white	low to medium	airy texture; some evergreen foliage; summer bloom
Hemerocallis	daylily	orange, red, yellow	medium	tough perennials with strap-shaped foliage; spring and summer bloom
Kniphofia	red hot poker	cream, orange, red, yellow	tall	striking racemes; sometimes evergreen foliage; late summer and autumn bloom
Lavandula	lavender	blue, green, mauve, purple, white	medium	evergreen foliage, often blue or silver; summer and autumn bloom
Lilium	lily	orange, peach, pink, purple, red, yellow, white	medium to tall	bulbs with often fragrant flowers; summer bloom
Monarda	bee balm	blue, maroon, pink, purple, red	medium	casual, shaggy flowers; summer to early autumn bloom
Nepeta	catnip	blue, pink, purple	medium	spikes of long-lasting flowers; summer to autumn bloom

Botanical name	Common name	Colors	Height	Comments
Penstemon	beard tongue	blue, maroon, pink, purple, red, white	medium	summer bloom
Perovskia	Russian sage	blue, purple	medium	aromatic, gray-green foliage; late summer bloom
Phlox		blue, pink, white	low to medium	clusters of often fragrant flowers; summer bloom
Salvia	sage	blue, lilac, pink, purple, red, white	low to medium	herbaceous or evergreen perennials; often colorful foliage; summer bloom
Scabiosa	pincushion flower	blue, pink, mauve, purple	medium	summer bloom
Sedum	stonecrop	pink, red, maroon, purple, white	low to medium	both foliage and flowers are colored; summer to autumn bloom
Verbena bonariensis	verbena	purple	tall	loose airy flower heads with long-lasting color
Veronica	speedwell	blue, pink, white	medium	spikes of intense blue flowers; summer bloom
Grasses				
Carex	sedge	bronze, green, tan, yellow	low to medium	mostly evergreen; grow primarily for foliage
Pennisetum	fountain grass	cream, gold, maroon	medium	featherlike flower clusters in summer
Phormium	New Zealand flax	bronze, green, maroon, red	medium to tall	strong, straplike, often multicolored foliage
Stipa	feather grass	cream, gold	medium to tall	airy flowerheads

Botanical name	Common name	Colors	Height	Comments
Vines				
Clematis		blue, burgundy, maroon, mauve, pink, purple, yellow, white		perennial vines; evergreen types are too vigorous, choose patio types
Lathyrus	sweet peas	almost every color		annual vines bloom in spring and summer; often fragrant; clear away spent stems after bloom
Nasturtium		cream, bronze, orange, red, yellow		annual vines with long period of bloom
Annuals				
Consolida	larkspur	blue, purple, white	medium to tall	
Delphinium		blue, purple, white	medium to tall	striking spikes of often intense color
Digitalis	foxglove	bronze, pink, purple, white	medium tall	airy spikes for cottage garden look
Cosmos		pink, white	medium to tall	feathery texture and daisylike flowers
Cerinthe		blue	low to medium	grayish foliage, arching habit
Nicotiana alata	flowering tobacco	white	tall	striking, large-leafed plants; tubular fragrant flowers
Shrubs				
Buxus	boxwood		low to tall	provides evergreen structure and a dark green foliage for colourful roses
Cistus	rockrose	pink, white	medium	summer bloom
Cotinus	smokebush	maroon	tall	reddish-maroon foliage makes a good foil for pink roses
Spiraea		pink, white	tall	deciduous shrubs with summer bloom

Pink 'The Fairy' softens a bright planting featuring red and burgundy tones, including red fountain grass (*Pennisetum rubrum*) and cerise *Aster novae-angliae*, backed by burning bush (*Euonymus alatus*).

Tall, deep blue *Delphinium* 'Pacific Giant' adds structure and height to a mixed planting with pale pink *Rosa* 'Blush Noisette'.

The new millennial rose garden

Clematis and roses are natural companions, as demonstrated by this *Clematis viticella* twining through *Rosa* 'New Dawn'.

Salvias come in a wide range of blues, reds, and purples to match almost any color rose. Here *Salvia nemorosa* is paired with *Rosa* 'Eliza'.

Perennials and annuals with tinted foliage offer opportunities to match with rose blooms, as shown by this combination of *Rosa gallica* 'Versicolor' underplanted with a blue-leafed pink (*Dianthus*).

'The Fairy' can be grown as a standard rose with a single trunk, as shown here surrounded by a pool of lavender flower spikes.

The small flowers of the airy perennial *Gaura lindheimeri* 'Siskiyou Pink' mingle with the shrub rose 'Sweet Meidiland'.

areas, herbaceous perennials and bulbs will die back to the ground every year, which will make it easier to get close to your roses for winter pruning. Others, like lavender, develop a woody framework that you may need to trim to prevent the plants from becoming leggy.

Scented plants and those with aromatic foliage are good choices for a mixed rose planting, and they help to attract beneficial insects as well. Vines like clematis can snake through rose bushes, giving the effect of a plant with two completely different but complementary blooms. Beware, however,

of planting vigorous vines like evergreen clematis or honeysuckle, which will quickly envelop your roses.

When it comes to color matching, there are some obvious combinations that work well together, such as orange with blue, cream with lilac, and red with mauve, but be careful not to overdo it in a single border. Choose a few colors in a palette, then pick plants that are variations on the theme. And don't forget about foliage. You can play with different greens, grays, blues, and bronze-tinted foliage to pick up tones in rose flowers, as well as the blooms of other neighboring plants.

VERTICAL ROSE GROWING

A breathtaking display of roses in the garden can be achieved by planting climbing roses on structures such as a fence, pillars, or gazebos. Against a fence or trellis, creating lateral canes (training the canes horizontally

Evergreen shrubs give structure to a formal court-yard planting with white 'Flower Carpet' rose.

A painted trellis and archway is the support for 'Climbing Pinkie'.

instead of vertically) will encourage the climber to respond by blooming all along the cane, often providing such a floral abundance that it can be hard to see the green leaves. Just imagine your plain fence being transformed into a living wall of color throughout the growing season.

ROSES IN CONTAINERS

I am a big advocate for planting roses in containers, because many gardeners may only have a small space in a city garden or a patio outside a condominium. You may choose a particularly fragrant variety to enjoy up close or complement the bloom color with a plant of contrasting color in the container. There are lots of possibilities, and I have listed here and indicated in the directory some roses that are particularly suitable for container culture. Choose a sufficiently large pot and fill it with excellent quality potting soil. Keep the soil surface mulched to help preserve moisture and provide slow-release nutrients. Be careful about additional fertilization

True red 'Florentina' scrambles up a pillar.

'New Dawn' thrives climbing over a gateway.

By tying the long, flexible canes of a climber against a trellis or wire support, you encourage the plant to send out flowering shoots along the entire length of the cane.

because concentrated synthetic fertilizer can harm plants or encourage leafy growth at the expense of blooms. Choose slow-release organic feeds instead. The most important consideration is to be conscientious about the watering needs of container plants, as they are more subject to weather conditions than roses grown in the ground. Roses in containers are particularly sensitive to the heat and they may dry out quickly if not regularly watered.

A vintage tub overflows with long-lasting color in the form of 'Pink Flower Carpet' rose, *Fuchsia* 'Beacon', and mauve *Campanula* 'Blaue Clips'.

Some of my favorite disease-resistant roses for containers are

'Coral Drift'

'Cream Veranda'

'Crimson Meidiland'

'David Rockefeller's Golden Sparrow'

'Dolomiti'

'Ducher'

'Easy Does It'

'Escimo'

'First Crush'

'Flirt 2011'

'Flower Carpet Amber'

'Innocencia Vigorosa'

'Jane Bullock'

'Julia Child'

'Kew Gardens'

'KOSMOS'

'Lady Elsie May'

'La Marne'

'Larissa'

'Marie Daly'

'Marie-Luise Marjan'

'Miracle on the Hudson'

'Oso Easy Cherry Pie'

'Oso Happy Petit Pink'

'Out of Rosenheim'

'Peach Drift'

'Pink Drift'

'Pink Martini'

'Pink Pet'

'Purple Rain'

'Raspberry Kiss'

'Republic of Texas'

'Roemer's Hip Happy'

'Rosenstadt Freising'

'Roxy'

'Ruby Ice'

'Solero'

'Spice'

'Sweet Vigorosa'

'The Fairy'

'Thrive!'

'Tupelo Honey'

Roses for small spaces | Because roses provide such a long season of bloom, they can act as focal points in a small garden. If space is limited, just one or two choice plants can do the trick and provide splashes of color when other plants have come and gone. It is wise to pay attention to the mature sizes of any roses as you want them to grow into their own glory in your small garden—sizes are listed in the directory. A rose that will grow to 10 feet tall when mature might have seemed desirable in the container in the garden center, but will become a nuisance as it grows. Too often I have seen people plant climbing roses around their mailboxes. As desirable as this sounds, the exuberant nature of the rose can quickly consume the mailbox and surrounding territory. 'New Dawn' is a good example of a plant that I have seen grown around mailboxes by mistake—this is a rose that wants to grow 20 to 30 feet high! I have seen gardeners hacking away at these poor plants to "keep them in line" and not grow beyond the limits of a too-small support. With a smaller variety, the size and shape of the plant will complement the space and not overwhelm it, keeping both the gardener and the rose bush happy.

Regional rose growing

Did you know that the rose is America's national flower? The proclamation, over thirty years ago, was more emotional and symbolic than practical, however. In fact, "national" is a little too broad for roses. I once asked fifty rose-growers across North America and Europe, "In your experience, what is the list of top ten roses that you have grown in your garden and would recommend to others as the toughest and most disease-resistant?" Unsurprisingly, not one list looked like another. There were only a few roses that were duplicated on some lists, but not a single rose made it onto every list.

I know there are gardeners in all fifty states who want to grow roses, not to mention all ten Canadian provinces, and all across Europe. As gardeners, perhaps we can consider ourselves citizen scientists, because we are interested in the sun and moisture as it directly affects our own gardens and backyards. Yet it is pretty clear that wherever roses are grown there are many different climate zones. An area's position on the continent, especially its distance from oceans and mountains, its altitude, and its latitude

In a small patio area, 'Flower Carpet' in a container provides a focal point of pink bloom.

Roses, an American symbol

In 1986, Congress designated the rose as the National Floral Emblem of the United States. President Ronald Reagan issued a proclamation stating, "More often than any other flower, we hold the rose dear as the symbol of life and love and devotion, of beauty and eternity." I want to thank President Reagan for this floral declaration.

The proclamation tells us that our first president, George Washington, was a rose breeder, and the rose 'Mary Washington' that he bred and named after his mother is still grown today. The White House itself has a rose garden. To me, a public rose garden is almost a sacred place. I imagine myself standing on the edge of a cliff, looking out and emphatically declaring, "I pledge to protect this symbol of floral joy and of our nation." Some people would probably suggest that I should get serious. But please humor me. Hyperbole aside, the public rose garden is a great responsibility as well as a pleasure, and I believe many people share the view that the rose has a special place and even a personal connection with themselves and the history of our country.

all influence the climate. Roses that grow well in Arizona, for instance, are unlikely to do well in New York, because the arid desert conditions in Arizona mean that fungal diseases are less of a threat than in New York, where they are encouraged by the summer humidity.

There are also drastic differences in day length from south to north: in early December, much of North America receives eight hours of daylight, while Orlando, Florida, gets ten and a half hours. In Northern Europe, by contrast, day length may be as short as six hours, but in southern Italy a December day can last well over nine hours. Day length and temperature affect both dormancy and bloom. Hardiness is a particular issue for growers in northern and mountain climates. Professor Griffith J. Buck (1915–1991) at Iowa State University created the Griffith Buck collection of roses, primarily to survive the harsh prairie winters of the Midwest. They are indeed tough, resilient,

and disease-free plants. However, many of them are not fragrant. As always, breeding roses for one characteristic often leads to the loss of another.

REGIONAL ROSE PICKS

Here are some roses that do particularly well in certain climate regions.

The Southwest has a desert climate with extreme summer high temperatures. In the most extreme summer heat, plant growth may slow and bushes produce fewer, smaller blooms. Fortunately, roses grown in hot, dry conditions do not usually succumb to diseases like black spot and powdery mildew. However, foliage can get dry and dusty, which encourages insects, so the foliage sometimes benefits from an occasional spray of water from the hose. Drip irrigation is a good choice to keep roses watered, and they need a good deep watering several times a week. Prune in midwinter.

'Grand Amore'

'Julia Child'

Southwest

'Beverly'

'Fiji'

'Easter Basket'

'Garden of Roses'

'Grande Amore'

'Jasmina'

'Plum Perfect'

'Postillion'

Savannah'

'Thrive! Copper'

during hot, dry spells. In winter, you may need to force roses into dormancy by stripping the foliage.

West

'Beverly'

'Easter Basket'

'Escimo'

'Grande Amore'

'Julia Child'

'Michel Bras'

'Postillion'

'Solero'

The California coast has a mild Mediterranean climate. Along the coast, rain falls through the winter, but the summers are dry. Despite the warm summers, roses grown in fog belts can be more susceptible to diseases. Keep your bushes watered by using drip irrigation, increasing the frequency if needed

In the Southeast and southern Florida the climate ranges from warm to subtropical to tropical, often with year-round average temperatures in the mid-60°s F. The summer heat and humidity can encourage diseases, so gardeners in these regions must choose plants that can do well even in these conditions. Plants may not go fully dormant in winter. Supplemental water needs will depend on the extremes of heat and humidity during the growing season.

South
'Alister Stella Gray'
'Belinda's Dream'
'Blush Noisette'
'Caramella'
'Climbing Pinkie'
'Cubana'
'Dark Desire'
'Ducher'
'Innocencia Vigorosa'
'Kardinal Kolorscape'
'La Perla'
'Lion's Rose'
'Mandarin Ice'
'Marie Daly'
'Marie-Luise Marjan'
'Mutabilis'
'Nastarana'
'Peggy Martin Survivor'
'Pink Pet'
'Republic of Texas'
'Rêve d'Or'
'Tequila'
'The Fairy'

'Blush Noisette'

The Midwest, Central and Northern Plains, and Great Lakes have a humid continental climate with four strongly defined seasons. Winters are cold and snowy, and summers can be hot and humid. Those summer conditions can encourage disease, especially black spot, so it's most important to choose the disease-resistant plants recommended in the directory. Choose a location in the garden where your plants can get at least six hours of sun a day. Roses in these cold-winter areas will go fully dormant in winter, and in the coldest regions you may need to choose cold-hardy varieties. Prune in very early spring, before buds on the stems start to swell.

'Rosanna'

'Poseidon'

The Northeast and Atlantic states also have cold, snowy winters and hot summers with plenty of humidity, which can encourage leaf spot diseases. Gardens close to the coast may also have to contend with salt spray and winds, plus the occasional hurricane. Roses need to be tough and disease-resistant to survive. Prune in these regions in early spring, before the buds start to break dormancy and swell.

Midwest

'Alexandra Princesse de Luxembourg'

'Black Forest Rose'

'David Rockefeller's Golden Sparrow'

'Larissa'

'Lemon Fizz'

'Morning Magic'

'Out of Rosenheim'

'Pomponella'

'Raspberry Kiss'

'Rosanna'

'Roxy'

'Souvenir de Baden-Baden'

'Summer Memories'

'Thrive!'

Northeast

'Brothers Grimm'

'Cinderella'

'Crimson Meidiland'

'Knock Out'

'KOSMOS'

'Larissa'

'Oso Easy Cherry Pie'

'Peach Drift'

'Poseidon'

'Purple Rain'

'Rosanna'

'Ruby Ice'

'Thérèse Bugnet'

'Thrive! Lemon'

'Topolina'

'Wedding Bells'

Winters in the North and in the high mountains can be long and cold, and summers are short, so choose roses that are hardy enough to survive those USDA zone 2, 3, and 4 winter lows. Roses grown on their own roots typically survive extreme cold weather better than grafted plants. Many gardeners mulch their roses in winter to give extra protection from the cold. Choose a sunny planting site, with protection from strong winds.

'Cape Diamond'

North

'Cape Diamond'

'Carefree Beauty'

'Centennial'

'Flirt 2011'

'John Davis'

'Lady Elsie May'

'Lena'

'Lupo'

'Macy's Pride'

'My Girl'

'Ole'

'Oso Happy Petite Pink'

'Quietness'

'Stanwell Perpetual'

'Sweet Fragrance'

'Thérèse Bugnet'

'Winner's Circle'

'Yellow Brick Road'

'Yellow Submarine'

In the Pacific Northwest, summers are cool and dry while winters are wet and overcast. Roses must be disease-resistant to

'Pink Double Knock Out'

NEW REGIONAL ROSE TRIALS AND LOCAL INFORMATION

Increasingly, roses are trialed to see which ones do well within particular climatic regions. When local growers and retailers offer these regionally adapted roses to local customers, everybody wins. Retailers have repeat customers because the plants they are selling result in happy gardeners. The customers are content because the plants they bought and planted in their yards are successful.

The year 2014 saw the launch of twelve test sites for a new type of rose trial, the American Rose Trials for Sustainability (A.R.T.S.). I am the executive director for these trials, which were in the works for several years. The trials are designed to prove the worth of roses based on the varied climatic regions of the United States. A.R.T.S. has a mission to identify and promote the most regionally adapted, environmentally responsible roses that do not depend on repeated chemical applications. The testing conditions are low-input, so the most beautiful, winter-hardy, and disease- and pest-resistant rose cultivars will be those that do well with little intervention.

The trials are carefully designed so that

thrive in the coastal dampness, often without full sun, but they may also need supplemental watering during the dry summer months, preferably delivered by drip irrigation that keeps the foliage dry. Pruning should take place in late winter, before new spring growth starts to emerge.

Many gardeners in northern regions see the frost clip their roses when they are still in bloom, sometimes even in early autumn.

Sunbaked walls in southern climates provide almost perfect conditions for roses—many varieties love the heat.

Foggy conditions in places like San Francisco can encourage fungal diseases on roses that are not genetically equipped to resist them.

the results will be scientifically valid. The roses are placed in randomized blocks with no two varieties next to each other in more than one block. The rating system is rigorous so that trial results can be submitted for publication in peer-reviewed scientific journals. State-of-the-art laboratory screening for black spot will help to characterize the disease-resistance of the roses being trialed. Any breeder or nursery can enter roses into the trials, and winning roses will receive a designation as A.R.T.S. roses, which will help to guide gardeners' choices for the best roses for their regions.

In addition to climate and weather, variables that affect growing conditions include water, soils, and local microclimates. Along with finding out about regional trials, a good source of discovering what grows well in your area is to look around and see what other people are growing successfully and ask them if they need to spray their roses to control diseases. Give your local rose society a call and you are sure to find plenty of people that will be glad to talk with you about what roses do well in their garden without the application of chemicals. Visit a local park or botanical garden to see what roses they are growing sustainably. Attend some garden tours, go to a garden show, or just walk around your neighborhood and see who is out spraying their roses with a breathing mask on and who is happily cutting fragrant blooms with spot-free foliage for arrangements. Which would you prefer?

Rose trials

In a typical plant breeding program, individual plants or genotypes are identified that exhibit a particular trait, such as drought tolerance and pest or disease resistance. These unique individuals are then bred to produce new roses with those traits. However, the development of new rose varieties

can take ten or twenty years to complete. Additionally, rose breeding is mainly carried out either by amateurs or highly competitive companies whose genetic information is often proprietary and unpublished. The goal of the Earth-Kind rose program is not to recreate these breeding programs of hybridizers around the world, but rather to identify those truly special roses that combine beauty with proven durability in the landscape. The Earth-Kind rose testing program began in Texas at the Texas A&M University, and has required the support of researchers in biotechnology, molecular genetics, plant breeding, plant pathology, and entomology. According to Earth-Kind testimony, the research this movement is doing is believed to be the largest environmental scientific rose research project to be conducted in the United States.

The research was funded by the Texas Nursery & Landscape Association and the Houston Rose Society rather than by breeders. This is an effort to independently and scientifically identify roses that are disease-tolerant, cold-hardy, low maintenance, and which stand as beautiful landscape shrubs (even without blooms). Of course, the blooming capability of each rose is of importance to the research as well.

The Earth-Kind philosophy is based on the premise that it is possible to identify beautiful plants that tolerate harsh, low-maintenance environments without agricultural chemicals and with significant reductions in irrigation. In the Earth-Kind model, the use of pesticides of any kind is only ever a last resort. If a pesticide becomes absolutely necessary, the protocol is to use the most "earth-kind" or environmentally responsible product available.

In the spring of 2010, we planted the Northeast Earth-Kind trial site at the New York Botanical Garden. It is a particularly

Rose trials carefully follow plants to see how they perform over a period of several years.

exciting accomplishment, as it truly helped us to identify some of the best possible roses for our region. At the end of 2013, the thirty-two varieties planted in the Northeast trials received the final scores that had been recorded over their four-year period. Trials also took place elsewhere in New York, Connecticut, and Maine.

EARTH-KIND GARDENING

In addition to the roses that are identified as Earth-Kind, the program recommends some best practices that eliminate the need for any application of synthetic or organic fertilizers. According to the Earth-Kind data,these practices also allow for a 98 percent reduction in applications of fungicides, insecticides, and miticides. I encourage these practices for home gardeners as well.

Earth-Kind roses coupled with Earth-Kind landscape management practices provide an easy-to-understand, comprehensive approach that allows gardeners to provide maximum protection for the environment. Because it is a landscape system that offers substantial savings in time and product costs, it is one that has been adopted by public gardens, city park departments, and homeowners around the country.

As of the writing of this book, seven universities are participating in Earth-Kind Rose research studies under the Texas AgriLife Extension research umbrella: Texas A&M Commerce, the University of

Earth-Kind trials have taken place all over the United States, to test roses in a wide range of conditions.

Minnesota, the University of Wisconsin, Iowa State University, the University of Nebraska, Colorado State University, Louisiana State University, the University of Arkansas, and Kansas State University. Horticulture departments at two community colleges in Connecticut and Illinois are also conducting Earth-Kind research.

Public and botanic gardens are engaged as well, including the New York Botanical Garden, UMore Park in Minnesota, Boerner Botanical Gardens in Wisconsin, Fort Worth Botanic Gardens and the Dallas Arboretum, both in Texas, the Columbus Park of Roses in Ohio, the American Rose Center in Louisiana, and Yew Dell Gardens in Kentucky.

Earth-Kind demonstration and field trials have been conducted by many rose societies and by more than thirty Master Gardener groups in Texas alone. With this kind of rigorous regional testing, only the most healthy and vigorous roses receive the designation.

As my dear friend Gaye Hammond, past president of the Houston Rose Society and fellow Earth-Kind collaborator says, "If you can grow weeds, you can grow Earth-Kind roses."

Growing roses sustainably

Rosa chinensis

Choosing the right roses is the first step toward sustainability in the rose garden. But how you prepare for those roses and how you care for them after you have brought them home are both important if you want to keep your garden chemical-free. This may mean doing things differently from the ways you have tried to grow roses in the past. My own approach to rose growing has changed dramatically over the years. I now realize that I don't need to pour chemical fertilizers on the soil and spray toxic pesticides on the leaves of my plants. I no longer see pests and diseases as my enemies that must be defeated at all costs in order to achieve perfection in the rose garden.

Part of the joy of the new millennial rose is that it takes much less work—and no chemicals—to care for these roses in your garden. You may be surprised by the simplicity of my advice, but I can assure you that all the methods I suggest are tried and true. Pick the right rose for your garden, plant it properly, and care for it well, and you will be able to enjoy thriving, healthy roses without the need for a hazmat suit.

Buying roses

A bare root rose is exactly what it says—a dormant rose that is sold to you without a container and with the soil washed away from the roots. This is a common way of shipping roses because the weight of any soil adds to the cost of shipping. The roots are usually brown in color rather than the active, young white roots seen in container-grown plants.

A container rose is a plant in a pot full of soil that is actively growing, showing leaves, and has an active root system. Often these roses are blooming so that the consumer is enticed to buy them. If you purchase roses at a garden center or nursery, they are likely to be container roses that came from a green-house or growing facility. In contrast, bare root roses usually come from an indoor dark cooler where the roses are kept until the time of shipping.

Bare root roses are usually shipped in the cooler months because the plants are dormant at this time. Container plants can be shipped and planted all year round as long as the soil at the planting site is workable (not frozen) and the shipping environment is not too hot. Some companies do not ship plants in the extreme heat of summer simply because of the danger of a plant being exposed to too much heat and drying out in the shipping process.

Bare root roses are packaged and shipped during the dormant season.

What do roses need?

Over the years I have been asked time and again, "What do I do with my rose when I take it home? How do I plant it and care for it?" My answer has evolved over the years, and nowadays I like to stress that the main thing is to understand what roses need in order to thrive.

ROSES THRIVE IN SUN

Period. Roses are not shade plants like hostas or hydrangeas. Roses need a full sun environment, which means at least six hours of direct sun every day. The amount of sunshine you receive in your rose site is proportional to the number of rose flowers you will be rewarded with.

I often get asked if a rose can grow in shade. My simple and honest answer is yes, but roses are not shade-loving plants and a rose growing in the shade will more

Whose roots are whose?

The vast majority of new roses are grown on their own roots rather than grafted onto a rootstock. Own-root roses take a little more time to build up and become established than grafted roses, but they are always true to their variety, are more likely to survive hard winters and temperature fluctuations, and tend to be healthier garden plants less prone to pathogens. I prefer own-root roses to grafted plants.

than likely just be a green plant—and a very unhappy green plant at that. It is true that some roses are more shade-tolerant than others, but in general I recommend that you stick with the rule that roses love full sun.

ROSES NEED AIR—BUT HOW MUCH?

Conventional wisdom states that planting roses close together so there is little space between them increases the opportunity for disease and pest pressures. However, in the new millennial rose garden, this is less of an issue. Although we do space out roses when conducting trials so that we can carefully examine the roses, in the Peggy Rockefeller Rose Garden, I have packed roses together very tightly. Not only does this give you the opportunity to really play off different colors and flower shapes in your planting scheme, but you can grow more roses! With roses that have been bred to resist disease, leaving room for air circulation is less of a necessity than with older varieties.

ROSES THRIVE IN GOOD SOIL

A basic soil test is an easy and inexpensive way to get a good snapshot of your soil's composition, biological activity, nutrient levels, organic matter, and pH, among other things. Soil tests are available from private suppliers or at your local Cooperative Extension Office or other horticultural agency or society. Following the instructions, you dig up small amounts of soil from various parts of your garden and then mix the samples into a single batch. When you send in your soil, indicate that the crop you are planting is roses, and the resulting recommendations will help you to correct any nutrient deficiencies in the soil.

Some roses will thrive in a wide range of soils and pH ranges. But it is always worth getting a soil test in order to have a basic understanding of what is going on with the

Rose Stories

New opportunities

Many years ago, I worked at a nursery in Atlanta. One day a favorite customer walked in and said she wanted to talk about roses. Normally I'm delighted to talk about roses, but I told this customer I didn't think it would be a good idea. I knew her garden well; it was very shaded by large trees. The vast majority of her plant acquisitions had been shade-loving plants.

"Why are you interested in roses all of a sudden?" I asked her.

"Remember that storm a couple of nights ago?" she replied, "Well, I now have a sun garden!" It turns out that a fierce windstorm had taken down several trees in her yard and the surrounding neighborhood.

I expressed to her my sympathies for her loss. But she surprised me with her answer, "Except for the unwanted cleanup of debris, I'm not that heartbroken. Because with the trees gone I have the sunshine that will allow me to have the rose garden I've always wanted."

Fortunately, you don't need to rely on a windstorm to create the kind of conditions that roses require. Sometimes a little judicious tree trimming will do the trick!

Container-grown roses are usually leafed out or even in bloom at the time of purchase.

soil in your own backyard. No matter what the results of a soil test, I recommend that you prepare your planting bed by tilling or digging 3 inches of compost into the top few inches of soil. The compost will provide sufficient nutrients to the plants for the first year of development.

ROSES THRIVE WITH WATER (BUT NOT TOO MUCH)

I always supplement my roses the first year with water to help them get established. A drip system is an ideal way to irrigate roses because it delivers water slowly and deeply to the soil, doesn't oversaturate the soil with too much water, and gets the water deep into the soil to help the roots get well established. In contrast, surface watering encourages roots at the surface where they can dry out and even burn. So if you do water by hand, make sure to soak the soil thoroughly and deeply. Also be sure to water all the soil in the bed and not just at the base of individual plants.

The first year of a rose's growth is when the roots are getting established in the soil. Although it's good for a plant to spread its roots and build a strong root system as they search for water, for the first year it is also good to make sure the plant has what it needs to survive. After the first year, I water in extreme heat and drought conditions only because I want the rose plants to continue to establish strong root systems that extend and search for their own water. However, take care not to over-water in extreme heat and drought conditions. In very arid climates, you need to pay particular attention to the watering needs of the plants. When temperatures reach a certain point, roses (like many plants) shut down as a self-protective measure. During this period of self-dormancy in response to heat, the plant is not absorbing much water, so don't add so much water that you saturate the soil. It's important to know whether the plant is actively growing and taking up water or in a dormant stage not taking up water. By careful observation of the roses, you can see the difference. A layer of mulch also helps to retain moisture in the soil and reduce the need for repeated watering.

ROSES THRIVE WITH DRAINAGE

Roses do not thrive in wet soil conditions. To see if your soil drains well, dig a hole about a foot deep, roughly where you want to plant your rose, fill the hole with water from a garden hose, and then watch to see how quickly the water drains naturally from the hole. If water remains in the hole for more than thirty minutes, then your drainage is poor and it would be best to improve the soil texture. Generally, it is better to improve your native soil than to bring in bags of new garden soil or sand. You can amend the bed with organic material such as chopped leaves, composted manure, or compost. As the organic matter decomposes into humus it slowly releases nutrients that are absorbed by microbes and bacteria in the

Acid or alkaline?

A particular test value that is important for growing roses is the pH of your soil. Typically, soils are placed on a pH scale ranging from 1 to 14. Acidic soils have lower readings on the scale (vinegar, for instance, has a pH of about 3), and alkaline soils have higher readings (milk of magnesia has a pH of 10). Different nutrients are most available to plants at different soil pH levels. For roses, the best soil pH is between 6.0 and 6.5, which is slightly acidic. If a soil test shows that your soil is considerably higher or lower than this value, the soil testing agency will recommend specific amendments that you can apply in order to bring your soil closer to this ideal range.

Drip irrigation delivered directly to the soil ensures consistent and thorough watering.

soil. In heavier soils like clay, humus pushes the firmly packed particles apart, drainage is improved, and it is easier for plant roots to penetrate and for water to drain. In sandier soils, the humus takes up residence in the large pore spaces and retains moisture.

ROSES THRIVE WITH MULCH

After planting roses, I strongly advocate applying a 3-inch layer of mulch. This mulch will stay on top of the soil to preserve moisture in the soil and block weeds. I cannot emphasize this point strongly enough. Mulch that is left on the soil surface will slowly break down and continuously feed the soil—and hence your plants—with nutrients, and with humus that provides proper drainage and aeration of the root zone and holds sufficient moisture to reduce the need for supplemental irrigation. Check your mulch layer in the spring and fall to make sure it is maintained at 3 inches. If the mulch has decomposed down to only 1 inch, then add 2 inches of mulch on top of it. Layer by layer, just add mulch on top as needed. In desert climates, increase the mulch to 4 or 5 inches, with drip irrigation lines running through it.

The best mulch is whatever is easy and accessible for you. Hardwood is a good option, keeping in mind that large chunks will take longer to decompose than small ones.

Again, make it easy for yourself by giving the roses what they need. Putting a rose in less than ideal conditions is a way to invite problems. A rose that is happy in its conditions, with plenty of sunshine and healthy soils, is going to be naturally healthy and disease resistant in your garden, eliminating any need for harsh chemicals.

Planting roses

There are so many choices of soil amendments and fertilizers that it can be confusing to the gardener. Keep in mind that the purpose of this book is to steer you to roses that have good genes that enable them to thrive without being spoon-fed plant food. Whenever one of my horticulture instructors, whether in a class on botany or ornamental horticulture, hinted that they might have a secret mix or soil amendment for planting holes, we students would quickly write it down in the hope that if we added this potion to our planting hole, we too would have success.

Some rose growers believe in lots of fertilizers and amendments at planting time; I used to be one of them. However, over the years I have learned that soils are incredibly dynamic and alive. Mother Nature has much more experience than I'll ever bring with my bags of stuff, and I now think that the best thing I can do when it comes to preparing soil is to test the soil for any deficiencies, and add as much organic material as I can. In a forest no one is spreading fertilizer around the trees to help them grow. They are fed organically by the slow decomposition of fallen leaves and branches on the forest floor. This continuous cycle can also become your secret potion for growing great roses.

You can plant own-root roses any time the soil can be worked, but it's best to plant grafted roses in the spring. It is best if you can prepare the entire bed before planting, but if you are planting into an existing border, you will need to start by digging a planting hole. However, I will stress that it is best if you can improve the structure throughout the planting bed. If you dig a hole in clay soil, for instance, and throw in some compost, you are in danger of creating a "bathtub effect" where water does not drain well into the surrounding soil. A loose-textured soil is not only better for roses, but it is easier on the person who has to dig the hole.

Pruning roses

For the first year of growth, I do not recommend pruning your roses. This allows the plant to stretch out its branches and use its increasing leaf surface to build up food stores by photosynthesis. The rose can then use its energy to grow up and gain strength. You may even wish to leave the plant to grow for a further season before pruning, as this will provide you with the opportunity to observe how the rose is growing naturally. This takes patience, but it is worth it as you are working with the rose so that it can do what it does best—bloom—and then do it again year after year.

There are many books on pruning roses, some of which offer extremely complicated instructions for different classes and growth habits. Remember that I am here to convince you that growing roses does not require a degree in engineering. I have pruned thousands of roses and I've come to the conclusion that when it comes to cutting back roses—especially the disease-resistant, healthy selections in this book—less is often more.

In my opinion, the key to successful pruning is to try and discern the natural form of the plant. How does it want to grow in my garden? Is it tall and upright? Short and

Mulch is key to keeping roses healthy. It provides slow-release nutrients, blocks weeds, and conserves moisture in the soil.

spreading? I have too often seen rose gardeners take their pruners to a rose plant with little or no understanding how the rose wants to grow naturally. Use the growth habits and plant sizes given in the directory descriptions for each rose to help guide you. For example, some plants "break" new growth from the top. It is from these new shoots that flowers will develop. If you are constantly pruning your rose bush from the top, you may be cutting off the future flowers. At the same time do not to worry too much about pruning after the second year because it is hard to

make a fatal mistake. With roses, whatever you cut off will grow back and you can learn from every poor pruning cut you make.

Again, classes of roses are based on the commonality of the blooms, which doesn't necessarily mean that there are common growth habits within each class. That's why generalizing about pruning becomes difficult. Many different varieties of roses have very individual growth habits. It helps to talk with other rose growers as experience tells that there are some roses that don't like to be pruned. For instance, some classes of roses, such as Teas and Chinas, some shrubs, and many Bourbons have a very delicate branching structure. I do not prune these plants much as they tend to build up upon themselves year after year in the garden.

Another question to ask about pruning is how you want that rose to function in your landscape. Do you want it to grow tall? Would you prefer to keep it more compact so that it thrives in a mixed border? Keep in mind that the objective of pruning is not to stop growth, but to direct it where you want it to go and to encourage vigor in the plant to bloom again (and again). However, don't try to impose a completely different growth habit onto a rose than it naturally possesses. What I have learned over the years is that the plants know exactly how they want to grow and they will try to grow that way no matter what. I always like to say that no matter what you cut off on a rose you will get it back. By making pruning cuts on your roses, you are directing the plant to spend energy and grow again. Probably the best advice I can give is to make some cuts on your roses and observe the resulting growth after you have cut. Spend time observing them and learning how this growth emerges and develops. Again, the best pruning respects the natural growth pattern of the plant, offering gentle encouragement rather than wholesale revision.

Basic planting steps

① Amend the soil in the planting bed.

② Dig a hole slightly larger and deeper than the root ball or bare roots of the rose.

③ Add compost to the dirt removed from the planting hole at a ratio of ⅓ compost to ⅔ soil.

④ Prepare the hole and plant the rose:

For a container rose, backfill the hole with the soil and compost mixture to where the bottom of the pot would be if placed in the hole. Place the plant into the hole and fill in around the root ball with the remaining mixture. Tamp in well to remove any air pockets. For an own-root potted plant, the soil level should be even with the natural soil level in the bed.

For a bare root rose, create a small mound in the hole and spread the roots over the mound, then backfill with the soil and compost mixture. Tamp in well to remove any air pockets. The soil level should be even with the natural soil level in the bed.

⑤ Water in well.

⑥ Top with a 3-inch layer of mulch.

The A.R.T.S. and Earth-Kind trialing systems do not allow pruning or deadheading during the trial period. If a rose succeeds as a winning rose for the A.R.T.S. two-year trials, it will then be qualified for entry into the Earth-Kind trials, which are even longer (four years) without pruning. The idea is to eliminate varieties that have straggly growth and need to be constantly pruned to shape. Is it a beautiful plant naturally? Does it have a landscape value? Does the plant have self-cleaning ability? What is the general impression of presentation of the plant? These are qualities and questions that can only be determined by not pruning the roses. Of course, some roses look horrible and have an awkward shape when unpruned. I steer away from these varieties once I witness how they want to grow in my garden. With the new millennial rose garden, I want to introduce the idea of less pruning because the plants do not require it.

There are two different types of pruning for roses. The first is winter pruning, when the plants are dormant. The other is summer pruning, when plants are in active growth.

WINTER PRUNING

The overall objective of winter pruning is to cut back the plant to the main branches so that it will respond with new canes the following spring. It is also a chance to remove any damaged or dead wood, whether it is from borers or from canes that have been physically broken. Do not prune too early in the season, but let your roses go fully dormant. The timing will depend on your region and your first frost dates. In the Northeast, I generally tell gardeners to stop making any cuts on their roses in September so that the plants are not stimulated to push out new growth that may get damaged by upcoming frosts. If you allow hips to form, the plant will naturally send its energy back down to

Fungi feeder?

Another tool in the toolbox for a rose garden without chemicals is the use of mycorrhizal products or mycorrhizae. Mycorrhizae are a group of naturally occurring soil organisms. Myco means fungus and rhiza means root, so the term refers to the symbiotic relationship between the two. The fungi use the carbon produced by the plants to support their own functions, and in turn they help the plant to reach farther into the soil by creating an extensive network or web of fungal filaments called hyphae that function like extended root hairs.

Research shows that roses with healthy populations of mycorrhizae are more vigorous, with increased drought and disease resistance and the ability to take up more nutrients and water.

Mycorrhizal products can be worked into the soil when planting, transplanting, or to inoculate an entire bed before planting. Inoculated soils will actually improve year after year, so it's a sustainable product. Once the mycorrhizae are incorporated, it is best to leave the soil undisturbed so the network of hyphae can develop and flourish year to year. If you till, or work the soil, it will break the connections that the fungi have with the roots and the process will have to start all over. Just maintain that 3-inch layer of mulch on the top of the soil. Avoid synthetic plant foods, especially fast-acting liquids, because they can harm microbial activity in the soil and thus create fertilizer-dependent plants.

the roots where it will be stored for the winter months.

In the hottest parts of the South and Southwest, or in other hot climates, roses are more apt to go into some kind of dormancy or slowed growth in the summer months to protect themselves from the heat. They then bloom more into the winter months. In this case, "winter" pruning cuts can be made during the summer.

In mild-winter climates like San Francisco, where winter lows rarely fall below freezing, plants may slow their growth but not ever become fully dormant. Sometimes I have forced dormancy on a plant by physically removing its leaves. This allows the plant to rest during the "off" months of midwinter. You can then winter prune before the weather starts to warm up.

In most four-season climates, fall progresses and roses prepare for the cooler temps and shorter days by dropping their leaves. Once the plant is fully dormant (in colder climates this may be midautumn, but in warmer ones it can be much later) with the resulting bare structure, you can study its architecture and see how it wants to grow. I don't spend a lot of time worrying about "outward-facing buds," although I know there are many other rose books and gardeners who do. In my years of growing and maintaining thousands of roses in private and public gardens, I have found it is almost impossible to spend the time to look for outward-facing buds on thousands of dormant plants and get them pruned in time for spring.

MAKING PRUNING CUTS

The first step of pruning can be called "reducing the complication." For instance, if you have upper growth on the plant that is very complicated, with many branches that further subdivide the growth into even smaller branches and then into even smaller, find the spot below this group of branches and cut them off. The idea is to remove all the little, fiddly canes (the complicated ones) back to a larger, thicker cane (a simpler one).

Remember that when you cut, it will encourage the plant to send out growth in a similar direction again. By observing the way the plant grew in the first place to produce the blooms that it did, you can expect that the plant will want to give you similar growth the following season. Roses can be very friendly and generous in this way.

I worry less now than I once did about crossing canes, interior branches, and vase shapes. Although a nice vase shape can be a great starting point from which a plant grows in the spring, I believe that if you chose plants with good genes, particularly with disease resistance, you will worry less about keeping the plant open to encourage air circulation. I believe that the rigid rules about pruning this way were necessary to help the disease-prone roses that we grew up with have a better chance of not getting disease. If you do see crossing canes, however, it doesn't hurt to remove them as they can rub against each other and cause damage.

Remember that with winter pruning you are making a cut on the plant that could expose the interior of the canes to potential pests and diseases. It is a good practice to dab a little wood glue on the end of a cane after making a cut to seal off the end against insects that like to bore into the canes and lay eggs.

After your roses have gone fully dormant, you may want to reduce the height of some of the taller plants. I typically cut to waist height (no matter what height your waist is) to remove some of the weight from the top of the plant. This will help protect against any buildup of snow on the plant that might cause it to bend and crack under the weight. In warm-winter climates, this might not be necessary, but it can give a more uniform look to the garden in winter. I always caution that if a rose bush has successfully produced hips to leave them for winter interest in the garden and for food for the birds. Certain classes of roses such as species roses and Rugosas do a very good job of producing hips.

SUMMER PRUNING

During the growing season, you make cuts to encourage the plant to produce the desired blooms. Note that this is different from deadheading, which is the removal of spent blooms in order to make the plants look tidy. Usually, deadheading has little to do with encouraging a duplication of the bloom you have already experienced.

In contrast to pruning dormant plants, the growing season is dynamic, constantly changing and variable depending on heat, insects, and weather in general. Roses like to grow when it's hot, so any pruning done during this season will encourage immediate regrowth. And remember that whatever you cut off on a rose you will get back. So if you are growing a long-stem Hybrid Tea for cutting, once the bloom has finished you can prune down the long stem to tell the rose to give you another rose on a long stem. If you only cut to the first set of six leaves as is commonly instructed, then you are only going to get the growth back from that very set of six leaves, and the bloom size will be proportionally smaller.

A lot of my summer pruning is dictated by the size of the blooms and the width of the canes, and by observing the branching structure produced by the plant to give me those blooms. Generally, a rose will produce a replacement cane the same thickness as (or less than, but never more) the one that was removed. If you observed that the plant's flowering came from a $1/2$-inch thick cane, then cut down to the same $1/2$-inch thickness to tell the plant to do it again. If you cut it down to $1/4$ inch, the resulting new growth will be at most $1/4$-inch thick and the flowers proportionally smaller. For example, if a Floribunda produced a nice cluster of blooms that came from a cane that was 2 feet long and at least $1/4$- to $1/2$-inch thick, I would cut the cane down to the same thickness and remove the branching that gave me the cluster of flowers.

The goal of deadheading is the removal of the spent bloom and thus the removal of any possibility that the plant can produce seed (hips). Because the plant has not produced seed, it will continue to bloom (in remontant varieties) with the goal of producing seed. By deadheading spent rose blooms, you can force a rose to continue flowering. However, to encourage your plants to repeat a consistent pattern of blooming, you cannot randomly snip off the old flowers. Instead, follow the instructions for summer pruning.

Rose care: feeding your soil

The compost you added to the planting bed will help get your roses off to a good start. Beyond year one, think about feeding the soil rather than your roses. If your soil is not

Summer pruning

healthy then your roses cannot be healthy. Maintaining a 3-inch layer of mulch that will slowly decompose over time provides humus (food) continuously to your soil and in return to your plants.

This is worth repeating. Slow decomposition is key. This gradual breakdown of organic matter allows for consistent and complete fertilization for the plant. The Earth-Kind data shows that in fifteen years of Earth-Kind research and field trials, a mulch of raw wood chips, preferably with some leaf tissue included as well, that remains on the surface and is not mixed into the soil works very well on roses and a wide range of other ornamentals. However, it's critical that the mulch remains on the surface and is not mixed into the soil. Many books on horticulture and plant care still warn that raw wood chips will rob the plants of nitrogen. This is what I was taught in my horticulture classes. However, as I have worked with the Earth-Kind movement, I have witnessed, based on their many years of trials across many states, this nitrogen depletion is simply not true with roses (or ornamentals in general) as long as the pieces of wood are ⅜ inch or larger and remain on the surface of the soil.

Synthetic fertilizers can be complicated and confusing. Which one is the best? Honestly, I have spent many hours and dollars trying all of the latest and greatest fertilizers available. I was addicted to the process of wanting to supplement my roses with the next best thing. If you do choose to supplement with fertilizers, I would most definitely recommend going with an organic product, that is, one derived from natural sources rather than being synthetically manufactured. Some good choices include blood and bone meal, chicken pellets, liquid fish fertilizer, compost tea, seaweed, worm castings, and kelp-based products.

Rose care: diseases

When we talk about rose diseases, we are usually talking about black spot, the most serious and often devastating disease of outdoor roses worldwide. It is caused by the fungus *Diplocarpon rosae*. Roses vary widely in their susceptibility to black spot, with popular Hybrid Tea cultivars typically being the most susceptible. Control measures require repeated sprays with fungicides, often as frequently as once each week from the first flush of growth in the spring until the first hard frost in the fall. The repeated and profuse use of fungicides at this level not only increases the cost to the consumer, but also can be potentially hazardous to the environment. It also places selection pressure on the fungus and can lead to the development of acquired resistance to different chemistries and pathogen populations.

There are a few other diseases that also affect roses. Rust is primarily prevalent in West Coast gardens, and it favors climates that have cool, moist weather conditions. It can be commonly recognized by the rust-colored, reddish-orange dots on the leaves.

Downy mildew is also common in coastal areas and can be easily recognized by purple blotches on the canes of the plants—I often tell people that they look like bruises on the stems. These lesions start off as yellow areas and mature rapidly into a brownish-purple color. Spots on the leaves can also be seen but these differ from black spot because the latter are circular, whereas mildew spots are irregular, almost trapezoidal (angular or blocklike). Downy mildew thrives in cool, moist conditions and can have a serious effect on susceptible plants.

Powdery mildew is very aptly named, as this particular mildew looks like a white, powdery coating over the buds and leaves. Favorable conditions for this mildew are the warmer days but cool nights, high humidity, and moderate temperatures. Severe infestation can cause stunting and distortion of plants.

Earth-Kind research evaluates each rose cultivar tolerance or resistance to common rose diseases. In order for a cultivar to be considered for Earth-Kind designation, it cannot get black spot more than once a year. And, if it does develop black spot, no more than 25 percent of the plant can be infected.

It is interesting to note that some varieties of roses have a tendency so that if they do get some black spot on their leaves, they don't defoliate. In other words, the genetics of the plant are strong enough to keep the leaves on the plant even with disease. The Earth-Kind designation does not mean that a cultivar never gets black spot. It means that if in controlled and adverse conditions the cultivar contracted black spot, the disease had minimal impact on the foliage and overall health of the roses.

I should mention that most gardeners and growers make a distinction between

Black spot is the bane of rose growers, but with new varieties, unsightly blotches on the leaves can be a thing of the past.

Rust causes orange patches on the leaves; fortunately it is the least prevalent of rose diseases and most modern varieties are resistant to it.

synthetic and organic products for controlling fungal diseases. When we use the word *chemical*, we are referring to synthetic compounds that are made in factories. Organic controls, on the other hand, are elements or compounds that originate in nature. Copper, sulfur, horticultural oil, neem, potassium silicate, and some anti-transpirants are used to control diseases on many plants, including roses. In addition, home remedies such as mouthwash and milk have shown some promise in providing disease control. I realize it is every gardener's personal choice whether or not to use such controls in the garden. However, the roses in this directory have been chosen because they do not require any kind of fungicidal spray in order to grow healthy, disease-free foliage. However, they may be more or less susceptible depending on your

climate and conditions. The regional plant recommendations will further guide you to the best roses for your garden.

Rose care: pests

I confess that I used to be a rose gardener who had an arsenal of every pesticide known to humans. I had a garden maintenance business in Atlanta and my clients depended on me to keep their roses perfect in every way, including protecting them from any and all insect damage. Now, after many years of growing roses, I approach things with a much different frame of mind. I have mentioned that I do understand that the right chemical at the right time can help solve a lot of problems. But I also understand that if I wipe out certain bad insects in my garden, I can also wipe out good insects that were dependent on the pest, either by killing them with the treatment or by taking away their food supply. What I try to do now—some times more successfully than others—is to promote a balance in my garden. I think it's okay to have some pest insects around as long as they are not consuming my plants into obliteration. I look to achieve a healthy balance of the good and bad in my roses.

WHAT IS IPM?

The Integrated Pest Management (IPM) approach recognizes that eradicating a population of pests in the garden will affect not only targeted pests but beneficial organisms as well. The goal with IPM is to manage pests rather than eliminate them, while at the same time exerting minimal impact on the environment. IPM uses biological, cultural, mechanical, and chemical controls to keep pest populations at tolerable levels. Remember above all that if a plant is repeatedly infested and unable to resist insect pressure then it is not meant for your garden.

The first step is to choose, plant, and care for roses so that they are as healthy as possible, which makes them more able to resist insect pressure. There are some roses that seem to resist pests.

To control pests, the gardener must be vigilant. It helps to learn to identify common pests (and beneficials) so that infestations can be nipped in the bud. Signs like rolled leaves or webbing can indicate concealed pests that are hard to see. The telltale signs of leafcutter bees are different from caterpillar damage, for instance (and those bees don't really do much harm). You may have to go out at night with a flashlight to look for weevils or slugs. When pruning, inspect canes for hollow or blackened centers that can mean cane borer infection.

Hand-control of insects is very possible in a home garden. Picking off beetles, blasting off aphids, and scraping scale insects of

Rose Stories

The problem of pests

Because I travel all over the country and lecture about roses, I get to meet all sorts of wonderful people who share my passion and are willing to spend an hour or so listening to me ramble on about roses.

I have learned that some gardeners are really plagued by the dreaded black spot fungus and others have no problem with it at all. I have grown roses in many harsh regions with high humidity that are prime for fungal diseases like black spot and mildew. However, I have rose-growing friends in California who claim that they don't have black spot. That makes me want to congratulate them on their growing conditions and yet throw tomatoes at them at the same time.

I always include a Q&A at the end of each lecture. It's typically friendly and fun and a lot of great information is exchanged. But invariably, there is always one person in the crowd who asks a question that goes something like this, "What do you do about Japanese beetles?"

There is a pause in the room. It's as though everyone suddenly forgets everything I had to say about disease resistance. If my answer includes anything at all that would help them with their dreaded pests, I would be a hero! When I'm asked this question, I deflate a little, but I have learned over the years to quickly come back with some humor.

"My whole lecture has been about roses and disease resistance in roses, not insects. I can tell that you are a troublemaker, there's one in every crowd."

Everybody has a good laugh and lightheartedness fills the room. But the question doesn't go away. Everyone actually wants to hear what I do about Japanese beetles.

"If I had an answer for Japanese beetles I would—" Well you know the end of that sentence: I'd be a wealthy man.

Flowering plants like alliums attract beneficial insects that will help to control insect pests on your roses.

Adult ladybugs on your roses are a good sign. It means you have help with the aphid patrol.

Adult syrphid flies are attracted to many plants from the daisy family; their larval offspring can consume hundreds of aphids.

Ladybug larvae look like little alligators and they happily munch on aphids.

rose canes can all help to prevent outbreaks from spreading. Sticky traps may also be helpful for some pests.

You can introduce or encourage beneficial predators in your garden. There are various ways to do this, from planting certain insectivore perennials, annuals, and shrubs to introducing predator insects and spiders to encouraging birds, bats, and other pest-eating creatures. Beneficial nematodes can be applied to the soil.

Here are some nectar- and pollen-rich flowers for your garden that will attract beneficial insects and birds throughout the growing season to aid in your pest control efforts. There are many other flowering and

Insectivore companions

Botanical name	Common name	Blooming season
Achillea filipendulina	Yarrow	summer through autumn
Allium		summer
Anethum graveolens	dill	summer
Asclepias tuberosa	butterfly weed	summer
Aster		summer
Astrantia major	masterwort	summer
Coriandrum sativum	coriander, cilantro	summer through autumn
Cosmos		summer through autumn
Lavandula	lavender	summer
Limonium	sea lavender, statice	summer through autumn
Monarda fistulosa	wild bergamot	summer
Penstemon		late spring through autumn
Petroselinum crispum	parsley	summer
Phacelia tanacetifolia	phacelia	late spring to early summer
Rudbeckia fulgida	gloriosa daisy	late summer through autumn
Sedum	stonecrop	summer
Tanacetum parthenium	feverfew	summer through early autumn
Tanacetum vulgare	tansy	late summer through fall
Thymus	thyme	summer
Veronica spicata	spike speedwell	summer

shelter plants that also encourage predators. Consult a local wildlife organization or native plant group for suggestions for your region.

PREDATORY INSECTS

These insects will all feast on pests in your garden.

Lacewings | Green lacewings in the genera *Chrysopa* and *Chrysoperla* are common natural enemies of aphids and other soft-bodied insects. The gray-green to brown alligator-shaped larvae feed on pests. The green lacy-winged adults feed on honeydew.

Ladybugs (lady beetles) | Many different red and black ladybug species are predators of aphids; the most common is the convergent ladybug, *Hippodamia convergens*. Another common species in the garden is the multicolored harlequin ladybug, *Harmonia axyridis*. Look for the black, alligator-shaped larva with orange dots and the oblong, yellow eggs that are laid on end in groups. Release ladybugs in the evening or by placing beetles on the base of the plants. Spray the beetles and the plants with water spray before release. Many will fly away, but those that remain and lay eggs will provide excellent aphid control.

Leatherwings or soldier beetles | These moderate to large-sized beetles have dark wings and orange or red heads and thoraxes. They feed on aphids and are very common on roses. Many people mistake them for pests, but they are predaceous both as adults and larvae (in the soil). Sometimes they leave dark splotches of excrement on leaves.

Minute pirate bugs | Minute pirate bugs, *Orius tristicolor*, are tiny insects with black and white markings as adults. They

Parasitic wasps lay their eggs in soft-bodied insects like aphids.

Spiders might not be welcome in your house, but in your garden they are efficient predators of many pest species.

are often among the first predators to appear in spring, and they feed on mites, insect and mite eggs, immature scales, and thrips.

Nematodes | Although some nematodes can cause damage to plants, there are also those that are beneficial. They are microscopic worms that live in the soil and can control many soil-dwelling insects. They must be mixed with water and applied soon after purchase.

Parasitic wasps | Tiny parasitic wasps lay their eggs within pests or their larvae. The developing wasp rapidly immobilizes the insects and turns it into a "mummy." The parasite pupates within the mummy and then cuts a neat round hole and emerges as a full-grown wasp. Once you see one mummy in a pest colony, you are likely to see more.

Pirate bugs and damsel bugs | Pirate and damsel bugs are natural predators of thrips as well as spider mites, caterpillars, and leafhoppers. They can be encouraged to visit the garden by the presence of flowering plants.

Predatory mites | Predatory mites feed on spider mites and can be distinguished from them by the absence of the two spots on either side of the body, their pear shape, and their more active habits. Compared to the plant-feeding species of mites that remain in one location feeding, predatory mites move rapidly around the leaf when looking for prey. Because they are so small, a hand lens is helpful in viewing them.

Predatory thrips | Six-spotted thrips feed on spider mites and are not plant feeders. Western flower thrips are plant feeders but also feed on spider mites.

Parasitic predators

When you see a pest insect on your precious roses, do you instinctively reach out to crush it with a gloved finger or flick it in the general direction of your chickens? If so, you might want to hesitate. If you see a beetle with little white dots on it and you kill it, you are also eliminating the next generation of parasitic insects that are growing inside those white dots. A parasitized beetle is unlikely to do any further damage to your plants, so a better option is to leave them so that the eggs can hatch and develop. This kind of observant monitoring of your roses is a way to ensure sustainability in the garden. Get up close and personal with your roses, not just to admire them but also to be on the lookout for the beneficials that are helping you to keep the garden chemical-free.

Spiders | All spiders are predators and many contribute significantly to biological control. Encourage them by not disturbing their webs.

Syrphid flies | Syrphids, sometimes called flower flies or hover flies, are important predators of aphids and very common on roses. Syrphids superficially resemble wasps, feed on nectar and pollen before reproducing, and are often seen hovering above flowers. Larvae, often found within aphid colonies, are legless and look like maggots.

Tachinid flies | These resemble houseflies but they live outdoors and feed chiefly on nectar, honeydew, and occasionally pollen. Tachinid flies over-winter as pupae in the long-dead shells of their beetle hosts, then emerge as adult flies during early June to lay their eggs, which can be seen on Japanese beetles as small, whitish dots glued behind the beetle heads. Attract them to your garden by growing plants with umbel-type flowers, including carrots, cilantro, and dill.

Aphids are common pests of roses and many other garden plants, feeding especially on soft new growth and flower buds.

INSECT PESTS

I will attempt here to address the most common rose pest problems. This is a challenge, because there are many variables and so many good insects look like the bad insects and vice versa. But generally speaking, here are some common rose pests that you may (depending on where you live) encounter in your garden and some ideas about how to deal with them in a sustainable way.

Aphids | These are very small, soft-bodied insects that can often be found in abundance smothering new tender growth or a new bud. As they suck the sap from the plant, they damage the leaves and flowers of the plant. Aphids are very easy to control without harsh chemicals.

Aphid control measures

Plant garlic and onions; the odor may drive aphids away.

Encourage ladybugs, parasitic wasps, big-eyed bugs, damsel bugs, green lacewings, minute pirate bugs, and syrphid flies.

Small birds like chickadees and wrens may be helpful in reducing aphids.

Monitor your plants, especially the new growth, for the first sign of aphids and then crush them with your fingers or knock them off your roses with a spray of water from a garden hose. Aphids multiply rapidly by parthenogenesis, so continue this process each day until they no longer return.

Mix garlic and water together and spray on the plants to act as a deterrent.

Mix up a home remedy spray of vegetable or mineral oil, water, and dish soap. This mixture sprayed on the aphids will suffocate them. Do not spray in bright sunlight or heat as the liquid can act as a prism and burn the leaves.

Spray with an organic insecticide or insecticidal soap that attacks the cell membranes of the aphids and causes their soft bodies to weaken and collapse.

Spray plants with a dormant oil spray, but do not spray when it is hot as you can burn the leaves.

Japanese beetles | These are hard-shelled, metallic-green, black, and gold beetles that can cause extensive damage to rose leaves and blooms. These beetles were first discovered in the United States in a nursery in New Jersey in 1916; they are believed to

Japanese beetles are tough adversaries, so enlist all the natural helpers you can in order to control them.

have arrived with some iris bulbs. Unfortunately as with many exotic pests, there are no natural predators that keep these pests at bay. Not only are Japanese beetles a threat to roses, they are devastating to a number of plant species including fruit trees, ornamental trees, shrubs, vines, crops, and turf grass. The beetle grubs exist in the soil, burrowing about 3 inches into the ground. The grubs spend about ten months of the year in larvae stage in the soil until they pupate and emerge from the ground to begin their feeding frenzy on your garden. The entire life cycle takes about one year.

A swarm of Japanese beetles can be difficult to control. My grandfather, Papa, and my grandmother loved their roses. Papa would laugh at the beetles and say it was a good daily therapy to go out into the garden and get rid of any frustration by "crushing the little suckers with my fingers." My grandmother used to enjoy this therapy as well. I remember both my grandmother and Papa flipping the beetles off into a bucket of soapy water so they would drown!

Leafcutter bees | It is unusual to see these insects at work, but they make their presence known by cutting perfectly circular holes at the edges of rose leaves. They use these leaf pieces to make egg partitions inside their burrows. The damage that is caused is strictly cosmetic and temporary, so there is no need to control them. In fact, leafcutter bees are important pollinators and are important native insects of the western United States.

Spider mites | Mites are very tiny relatives of spiders. They can be red, black, or brown in color. Mites pierce the underside of rose leaves and suck sap, causing the leaf to turn gray or bronze. To the naked eye, they look like tiny moving dots on the underside of the leaf, much like finely ground black or red pepper.

A fine web is a sign of heavy infestation. Mites reproduce rapidly, resulting in high populations in a short time. Mites flourish in crowded, stagnant gardens. I always look for spider mites when it is hot, dry, and dusty because they love those conditions. Plants under water stress are even more highly susceptible. As foliage quality declines on heavily infested plants, female mites catch wind currents and disperse to other plants. The use of chemical fertilizers can actually lead to an increase in spider mites by increasing the level of nitrogen in leaves.

Spider mite control measures

A high-pressure washing with water from a garden hose directed to the undersides of the leaves every two or three days can safely manage mites. Also spray or mulch pathways to keep down dust. High mite populations decrease in late summer as the weather turns cooler and rains begin.

Spider mites have many natural enemies, including predatory mites, six-spotted thrips, spider mite destroyer ladybugs, minute pirate bugs, big-eyed bugs, and lacewings.

Spray with plant extracts formulated as acaricides can help to control spider mites. These include oils made with garlic extract, cinnamon, clove, mint, and rosemary.

Treat with insecticidal soap, or with insecticidal oil such as horticultural oil or plant-based oils such as canola, cottonseed, or neem oil. Don't use soaps or oils on water-stressed plants or when temperatures exceed 90°F. Oils and soaps must contact mites to kill them, so be sure to cover all the foliage, especially the underside of leaves, and repeat as needed.

Thrips | Thrips are extremely small, slender, brown-winged insects that usually live and feed inside rose flowers. Discolored or distorted plant tissue or black specks of feces around stippled leaf surfaces are signs of thrips. Most thrips range in color from translucent white or yellowish to dark brown or blackish, depending on the species and life stage. These pests seem especially attracted to lighter colored roses. Note that some thrips are not pests, but are either beneficial or innocuous. Look carefully for the insects themselves to be certain that pest thrips are present and the cause of damage before taking control action.

Thrips control measures

Trim away dead branches to keep a plant healthy; improper or shallow cuts promote new growth that the thrips like to feed on. Proper feeding and watering will also benefit plant health.

Many thrips thrive on weeds, so destroying or removing their natural habitat can cause them to move onto the cultivated plants. If you can stand a few weeds around, the thrips might be more attracted to them than your roses!

Beneficial thrip species include black hunter thrips and six-spotted thrips. Predatory mites, lacewings, pirate bugs, damsel bugs, and nematodes can also help to control thrips.

Spray with insecticidal soaps or horticultural oil.

Rose midges | Midges are tiny flies that lay eggs in the new growth of buds and shoots of roses. The larvae feed and cause bent buds or withering stem; the tips turning black

Scale insects have hard bodies and adhere to plant stems but are easily scraped off if you find them before they have a chance to multiply.

as if they had been burnt. The presence of midge is generally correlated with the heat of midsummer and then followed by a second season in late summer and early autumn.

Rose midge control measures

The best non-chemical control consists of pruning out infected buds, but by the time you see the damage, the rose midge could have already come and gone.

Because the larvae fall to the soil to pupate, an effective control is to place weed barrier fabric under the plants to catch the larvae and prevent them from entering the soil. A thick layer of mulch may also help to keep midges at bay.

Scale | Adult scale insects are generally small and immobile, with no visible legs. They secrete a waxy covering, making some appear white and cottony while others appear like white, yellow, or dark bumps. This waxy covering or scale protects adult scale insects from many insecticides. Their immature forms, called crawlers, are susceptible, however.

Several species of scale are pests of roses, but rose scale (*Aulacaspis rosae*) is one of the most serious. Female rose scales are round, gray to white and about $1/16$-inch long. Males are smaller, longer, and white. Rose scales are usually found on rose canes where they feed on sap with their piercing mouthparts. With a heavy infestation, rose scale can cause cane decline or twig dieback.

Scale control measures

With light infestations, scale can be scraped off by hand and destroyed.

Prune out and destroy heavily infested canes.

Ladybugs and parasitic wasps feed on scale insects.

Horticultural oils penetrate the waxy coating on scale and smother them. Apply at higher concentrations of 3 to 4 percent during the dormant season to penetrate the thick waxy covers of the overwintering adults. Apply at lower rates of 1 to 2 percent during the spring to target the crawlers and the newly settled scales with thin waxy covers.

Leafhoppers | Adult rose leafhoppers (*Edwardsiana rosae*) vary in color from white to gray to yellow to green. They are wedge-shaped and between $1/4$ and $1/2$ inch long. When a plant is disturbed, they hop or fly away quickly. The adult female deposits eggs within the bark of rose canes in the fall. In the spring the young nymphs (immature forms that resemble adults but are wingless) emerge from the cane. The wounds that remain in the bark as they emerge, as well as wounds made during egg laying, can provide openings for stem canker-causing fungal pathogens to enter.

Nymphs and adult leafhoppers feed on the undersides of leaves, sucking plant sap. Their feeding causes small white dots on the upper surface of the leaf. Damaged leaves may drop prematurely. Between feeding by the nymphs and adults, and egg laying by adult females, a severely infested rose bush may be killed.

Rose sawfly larvae love to chomp on your rose leaves but they are easy to spot and dispose of.

Leafhopper control measures

During winter pruning, look for dark, purple, pimple-like spots on the bark that indicate the presence of eggs. Prune any infected canes.

Natural enemies of rose leafhoppers include damsel bugs and assassin bugs.

Rose slugs (sawflies) | Rose slugs are the larvae of sawflies, non-stinging members of the wasp family. Three species are rose pests: the rose slug (*Endelomyia aethiops*), the bristly rose slug (*Cladius difformis*), and the curled rose sawfly (*Allantus cinctus*). The larvae of some sawfly species are hairy and often mistaken for caterpillars. Others appear wet and shiny, superficially resembling slugs. The larvae generally reach about $1/2$ to $3/4$ inch in length. Sawflies get their common name from the female's tubular egg-laying organ, which unfolds like a jack-knife. It functions like a saw blade, allowing her to cut into stems or foliage and deposit her eggs. It's easy to mistake this feature for a stinger, but sawflies are harmless to people and pets.

Generally rose slugs feed at night. Depending on the species, young rose slugs feed on the upper or lower surfaces of leaves between veins, leaving windows of translucent tissue that turns brown. Regular inspection of your roses is important because feeding typically progresses quickly. In addition, with their coloring, they can be very difficult to spot on leaves. However, damage caused by rose slugs is rarely extensive.

Rose slug control measures

Handpick rose slugs off leaves.

Spray with a blast of water from the hose. Once dislodged, the pests cannot climb back onto the plant.

Spray with insecticidal soap or horticultural oil.

The chemical-free rose directory
150 disease-resistant roses

The roses outlined here are specifically chosen for this book. They have been proven in my experience to have the genes to resist disease and fit into a sustainable, chemical-free garden. These are the new millennial roses that will allow you to successfully grow these beautiful plants.

I realize that much of our appreciation of roses—and other plants—is subjective. Fragrance is a perfect example of this. Someone may smell a rose and make a comment on how very special the fragrance is. Maybe it reminds them of years past or a special place in their life. But another person may not have the same reaction. In general, I'm not going to tell you how to define the smell of a rose. I believe that we experience scents individually.

Another example of subjectivity is color. I may call a rose the color orange, and you may call it apricot. We all have our favorite colors and others that leave us cold. Fortunately, with roses you can get a pretty wide rainbow of colors. You may not quite like that shade of orange but you are crazy about apricot? There will be a rose in this directory for you. You will eventually find your favorites among favorites. Maybe you will fall in love with the orange one because of its other wonderful qualities. I guarantee that your favorites will be different from mine. But in order to get there you have to start growing roses, and this is a list to get you started.

With each rose description I have included some companion plants. These are roses that work well together based on my gardening experiences. All the rose companions in this menu are pictured in this directory, so if you like one, it will help you to pick other varieties to mix and match yourself.

Rating roses

All the roses in this directory have been assessed over several years in many different regions. I have assessed them all myself at the Peggy Rockefeller Rose Garden and in my own home garden. The results are given in the ratings for each rose which reflect the three qualities that are important to every sustainable rose gardener: disease resistance, bloom, and fragrance.

Disease resistance

55-60 / Superior resistance.
Plant is fully foliated and shows no signs of disease.

50-54 / Excellent resistance.
Plant shows up to 10 percent defoliation due to disease.

45-49 / Very good resistance.
Plant shows 10–15 percent defoliation due to disease.

40-45 / Good resistance.

Bloom

21-30 / Profuse number of blooms.

11-20 / Significant number of blooms.

0-10 / Good number of blooms.

Fragrance

7-10 / Very strong fragrance.

4-6 / Strong fragrance.

2-3 / Mild fragrance.

1 / Faint fragrance.

0 / No fragrance.

Overall rating
(Disease resistance + Bloom + Fragrance)

90-100 / Superior rose.

80-89 / Excellent rose.

70-79 / Very good rose.

60-69 / Good rose.

This rating system allows you to choose roses based on what is most important to you. If you live in a very disease-prone area with high summer humidity, you might only choose those roses with superior scores in the first category. If you live in a region with less disease pressure, you can feel free to try the most fragrant roses even if their disease resistance is slightly lower. I should stress, however, that all of these roses are chosen for this book primarily because of their ability to withstand disease. There are no bad choices.

Roses with the lowest overall rating (60–69) may have incredible disease resistance but may not bloom very much, or they are not fragrant. In general, Hybrid Teas score lower because they bloom only one flower per stem and don't give the same abundant season of color as a shrub rose.

Many breeders enter their roses into competitions and trials at the national and international level. I have listed the results of some of these competitions, but it is by no means exhaustive. For instance, you may see the initials ADR, which stands for Approved German Rose (*Allgemeine Deutsche Rosenneuheitenprüfung*) and is the result of a trial that is considered the toughest review in the world. It is helpful for a gardener to see that a rose has done well in places as far-flung as Austria or Australia, but a rose that has received no or few awards may still be a worthy choice for your garden.

Rose Stories

Roses for Rose

One gorgeous spring afternoon I was at the Peggy Rockefeller Rose Garden when I saw a gentleman pushing a lady in a wheelchair. I walked up to both of them and explained that I could show them a route to the garden that wouldn't involve steps. The gentleman took me aside and gently told me that the lady in the wheelchair was his wife, and she had come to fulfill her dying wish—to see the roses on this perfect day. I realized that my mission at this moment was to help this lady experience the garden one last time. By the way, her name was Rose.

Rose was a little too fragile to move around much so I took her on bird's eye virtual tour of each and every rose in the garden. I told her stories of roses and described them the best I could with phrases like, "This one is like velvet," "It is a brilliant, luscious red that knocks your socks off," "My grandmother, who you would have loved, had this one in her garden." In return, Rose told me stories of her own roses in her garden and how much she loved them over the years.

Other moments, Rose and I just sat there in silence looking at what seemed to be an unending display of joy in the thousands of blooms in the garden. It was a true gift. Although I never saw Rose or her husband again, I can tell you without hesitation that her last wish came true.

The point of this story is that roses are so special and bring so much joy to so many people in so many different ways, they can be hard to describe. A rose that is perfect and special to one person might take on a different shape, form, color, fragrance, or altogether different experience for another.

I wish I could go through the garden with each and every reader and experience with you the joy of roses one on one, but keep in mind as you read through this directory that you will find your own favorites within it.

'Above and Beyond'

Disease resistance	58
Flowering	25
Fragrance	1
Total rating	84

Large-flowered climber

Climbing to about 10 feet

Soft apricot

This new, large-flowered climber is a very different rose for colder climates combining extreme cane hardiness, prolific and reliable spring flowering, warm flower color, and vigorous growth. Clusters of typically five or more orange flower buds per stem open into apricot-colored, semi-double to double flowers. The petal arrangement and the frilly nature of the blooms is unusual in a climber in this color. Plants bloom heavily in midspring to late spring, with sporadic repeat flowering in summer. With extreme cane hardiness, 'Above and Beyond' can serve as a dependable climber or a nice large freestanding shrub in colder regions.

Introduced in 2014 by Bailey Nurseries; hybridized by David Zlesak (ZLEEltonStrack).

Companions | The climbing habit of this rose means you can grow it up a fence, pillar, or similar structure where the color can really brighten an area of your garden. Complement the apricot colors with 'All the Rage', 'Cubana,' 'Darlow's Enigma', 'KOSMOS', 'Postillion', and 'Sunny Sky'.

'Alexandra Princesse de Luxembourg'

Shrub

Tall and arching or spreading to 5–7 feet

Lighter pink outer petals sometimes fading to creamy white

This is one impressive rose. Everyone who walks past stops to admire it and put their noses right in the middle of the bloom—and they are not disappointed. It is an upright grower with very robust and healthy canes that produce an abundance of repeating, full-bodied blooms with a nice fragrance. The flowers are scalloped on the edges and have a somewhat old-fashioned appearance, yet seem very fresh and modern. Give 'Alexandra Princesse de Luxembourg' room to grow, as plants can reach 7 feet once established. In warm climates, it tends to get even larger and you could perhaps train this rose as a climber.

Introduced in 2009 by Kordes (KORjuknei).

Companions | The lovely shade of pink mixes well with lots of color palettes and 'Alexandra Princesse de Luxembourg' is a great specimen plant in the mixed border. You can also pair it with pink roses, both the lighter 'Cinderella' and 'First Crush' and the darker 'Eliza' and 'Wedding Bells'. White roses such as 'KOSMOS' and 'Summer Memories' will bring out the lighter tones in the petals.

Disease resistance	51
Flowering	20
Fragrance	8
Total rating	79

'Alister Stella Gray'

Disease resistance	**60**
Flowering	**27**
Fragrance	**5**
Total rating	**92**

Noisette, Tea Noisette

Tall and arching or spreading to 8 feet

Light buttery yellow, fading to creamy white

Every time I walk by 'Alister Stella Gray' it causes me to pause and admire its elegance and beauty. This plant can grow to 7 or 8 feet and is often used as a climber. I grow it on a pillar, which allows the wonderfully fragrant clusters of blooms to be right at nose level. Each single bloom is small, subtly complex, and a wonderful muted color.

Introduced in 1894 by Alexander Hill Gray.

Companions | Because the color of 'Alister Stella Gray' fades with age, I enjoy combining it with other pastels, creams, and whites. It can be placed near the back of a mixed border with some complementary colors in front. Planting with other yellow roses will highlight the wonderful yellow bud color. Good choices would be 'KOSMOS', 'La Perla', and 'Summer Memories' (creamy white) or 'Sunny Sky' and 'Winter Sun' (both soft yellow).

Disease resistance	**60**
Flowering	**28**
Fragrance	**0**
Total rating	**88**

Shrub

Bushy growth to 4 feet

Shining apricot-coral

Belonging to the Easy Elegance collection, 'All the Rage' is aptly named. Over and over I have seen people stop and stare at the flowers and then look at the name and say, "Of course—it's all the rage!" Glossy foliage sets off a unique array of flower colors, which is like a sunset on a gorgeous summer day. In my experience, the blooms are typically either single or in clusters of two or three. The simple, open flower form reveals a wonderful set of stamens surrounded by halo of gold at the base of the petals, which then deepen in color toward their edges.

Introduced in 2007 by Bailey Nurseries; hybridized by Ping Lim.

Companions | The many shades of coral, apricot, and peach in 'All the Rage' allow for a lot of possible color combinations with other plants, but it looks especially delightful planted with yellows and pinks. Some good rose companions are 'Alister Stella Gray', 'Centennial', 'Eliza', 'Sunny Sky', and 'Winter Sun'.

'Awakening'

Disease resistance	55
Flowering	22
Fragrance	5
Total rating	82

Large-flowered climber

Can grow to 10–15 feet tall and wide

Light pink

'Awakening' isn't a newcomer in terms of recent genetic developments, but it is nonetheless worth mentioning. This rose is a sport of the popular climber 'New Dawn' with a more double-petaled (almost quarter-petaled) bloom than its parent. It also seems to be a better repeat bloomer. This is a really strong grower that can easily scramble over a fence where it can stretch itself as wide as it is tall. The canes are thick but easy enough for you to train on a structure. The glossy foliage is typically without disease, full and lush, providing a nice background for the softness of the color of the blooms.

Discovered in 1935 by Jan Bohm; also sold as 'Probuzeni'.

Companions | Plant 'Awakening' on a fence or structure and give it some room to stretch. With the softness and paleness of its blooms, it complements plants with other pastel colors. Try combining it with 'Thrive! Lavender' to introduce some purple tones, or with 'Peach Drift' for some soft peach and apricot shades. Any pink rose will also do, including 'Belinda's Dream', 'Carefree Beauty', or 'Cinderella'.

Climber

Bushy and climbing up to 8 feet

Brilliant orangey pink, with a prominent yellow eye

This is a rose to turn heads! The jazzy color and bright yellow centers of the blossoms will catch your eye and keep you staring as you try to discern what color it really is. Orange? Deep pink? The buds show some golden-yellow before they open to a brighter orange, creating an overall effect of orange, pink, and yellow. The flowers open flat and are simple in form revealing the contrasting yellow eye and batch of stamens. 'Bajazzo' is a healthy rose and I have never seen a spot on its foliage, which is a deep, glossy green. Its flexible growth habit lends itself to a position on a pillar, light post, mailbox, trellis, or small fence.

'Bajazzo' grows strongly in a range of climates, as evidenced by the medals it has won from many different regions. It has been in international rose competitions and won, among other awards, a silver medal and an Honorary Award at the Baden-Baden Rose Trials and the ADR Rose Award (both 2010).

Introduced in 2011 by Kordes (KORteheba).

Companions | Versatile 'Bajazzo' can be grown in a variety of ways. Use nearby yellows and golds to visually pick up the yellow center of the blooms. Try the roses 'David Rockefeller's Golden Sparrow', 'Golden Gate', 'Lemon Fizz', or 'Sunny Sky'.

Disease resistance	**58**
Flowering	**21**
Fragrance	**0**
Total rating	**79**

'Belinda's Dream'

Disease resistance	**57**
Flowering	**23**
Fragrance	**3**
Total rating	**83**

Shrub

Upright and bushy to 4–6 feet

Medium pink

Some of the roses in this directory have a proven propensity toward disease resistance, while others have been specifically hybridized toward disease resistance. Some roses fit into both categories, and 'Belinda's Dream' is one such rose. It has been around for a while, and has proven itself so well as a tough, disease-resistant variety that it has been given the Earth-Kind designation and it was awarded Best Established Shrub at the 2013 Biltmore International Rose Trials.

'Belinda's Dream' has the ever-popular classic flower form—high-centered, with doubled petals that unfold beautifully. It keeps producing these pink beauties all during the season. The only downfall that I can see with this rose is that the petals do not fall cleanly off the plant, but that is a small price to pay for such a dreamy rose.

Bred in 1988 Robert E. Basye.

Companions | 'Belinda's Dream' grows quite large if left unpruned, so it can act as a striking specimen plant in the mixed border or massed to form a flowering hedge. The color is easy and flexible to mix with roses that have peach and apricot tones, such as 'Blush Noisette', 'Pink Bassino', 'Quietness', 'Roemer's Hip Happy', or 'Savannah'. Flowering plants and shrubs with cream and white blooms are also a sure bet.

Disease resistance	54
Flowering	21
Fragrance	10
Total rating	**85**

Hybrid Tea

Upright and bushy to about 4 feet

Pink blend

I have been preaching about disease-resistant roses for many years now, but there is one thing I don't talk about as much, and that is fragrance. I find that the fragrance of a rose is a very subjective experience, but 'Beverly' is a rose that defies this general principle, with a fragrance that can accurately be described as fruity and strong. There are few Hybrid Tea roses in this book because they have been hybridized for the high-centered flower form and not necessarily for disease resistance. Once again, 'Beverly' is the exception to the rule. The flowers are pink, with gradients of lighter pink to almost silver. The overall effect is captivating, at least for anyone who likes high-centered roses with healthy foliage and fragrance.

The popularity of 'Beverly' is borne out by the numerous awards it has won, including the Fragrance Award in Baden-Baden 2008, the Fragrance Prize and Audience Prize at Nantes in 2010, the Scent Prize in Belfast in 2009, a gold medal at La Tacita 2010, and silver medals in Baden-Baden (2008), Echigo (2009), and The Hague (2012).

Introduced in 2007 by Kordes (KORpauvio).

Companions | The base pink and gradients of pink that are present in the blooms of 'Beverly' allow for great for great flexibility with color matching and contrasting. Any white or lighter pinks highlight this color scheme. Try the companion roses 'Alexandra Princesse de Luxembourg', 'Eliza', 'Larissa', 'Lion's Rose', or 'Macy's Pride'.

'Black Forest Rose'

Disease resistance	60
Flowering	23
Fragrance	0
Total rating	83

Floribunda

Upright growth to 3-4 feet

Red

The Black Forest in Germany is a setting for many of the Brothers Grimm fairy tales, cuckoo clocks, delicious Black Forest ham, even more delicious Black Forest cake, and now 'Black Forest Rose'—*kostlichsten aller*, the most delicious of all!

'Black Forest Rose' has only been around for a few years but it has already established itself as one of the most disease-resistant roses, which is exceptional for a red rose. I have yet to see any black spot come near this rose in my garden. The deep green, extremely healthy foliage is a beautiful foil for the flowers, but you may not be able to see the forest through the trees because this rose produces so many flowers that the foliage may be completely covered. From a distance, 'Black Forest Rose' looks like a big ball of beautifully ruffled red flowers. I have seen forty to fifty blooms per cluster. The flower clusters grow on long, sturdy stems and each individual flower is small, very rounded, and almost ball-shaped. The color is a nice, medium red, but on the reverse of the petals (which are presented so precisely in this ball shape) is a light color with some pinkish tints. On top of all that, the flower petals have a frilly edge, adding deep complexity of the texture of the bloom.

This rose is a forest of pleasure for any gardener to enjoy and it has won multiple medals at international rose competitions, including an ADR Rose Award (2010), a gold medal at the Kortrijk Rose Trials (2012), and First Prize at the La Tacita International Trials for new roses (2001), among others.

Introduced in 2010 by Kordes (KORschwill).

Companions | 'Black Forest Rose' is an excellent candidate for mass plantings in home gardens and public spaces. It makes a perfect hedge that provides a nice splash of red almost throughout the season. The red color and the slight reverse color of the petals are fun to match with a wide range of companions. Try 'Caramella', 'Darlow's Enigma', 'Sunny Sky', 'Tequila', 'Thanksgiving Rose', or 'Zaide'.

'Blush Noisette'

Noisette

Bushy, upright growth to 5–7 feet

White to lightest pink

Disease resistance	**56**
Flowering	**23**
Fragrance	**10**
Total rating	**89**

The Noisettes are certainly not new kids on the block, having originated in the early 1800s. This class of roses has proven itself in the landscape over two centuries, which is why have included a few Noisettes in this directory. They make wonderful additions to any chemical-free rose garden. Noisettes are known as southern plants because they thrive in hotter conditions, but I have successfully grown them in some northern states, depending on the individual growing conditions.

The individual blooms of 'Blush Noisette' are not large, but in typical Noisette fashion they are produced in huge clusters to give a wonderful floral display. Given that Noisettes also have wonderful fragrance, this rose adds charm and beauty to any garden.

Hybridized in 1814 by Philippe Noisette.

Companions | Noisettes can take up a lot of space, so 'Blush Noisette' can be a specimen plant in the border. I have also used it on a pillar where I can get close up to enjoy the fragrance. The colors of 'Blush Noisette' allow for a wide variety of combinations with other pastels, pinks, and whites. Try 'Eifelzauber', 'Darlow's Enigma', 'Peach Drift, 'Quietness', or 'Stanwell Perpetual'.

'Brilliant Veranda'

Disease resistance	52
Flowering	25
Fragrance	1
Total rating	78

Floribunda

Compact, bushy growth to 3 feet

Brilliant orange-red

Brilliant! This is another brightly colored rose to draw your attention. 'Brilliant Veranda' (sometimes sold as 'Brilliant Flower Circus') is well named, with bright orange-red blooms and beautiful, deep green and glossy foliage. The flowers are about 2–3 inches across and are borne in clusters, as are other roses in the Veranda series. The beautiful form of this plant proves its worth in the landscape, and on top of that it rewards you with cheerful color throughout the growing season. Since its introduction, the quality and health of 'Brilliant Veranda' has meant it has consistently received excellent scores in rose trials. Other fine Veranda roses include 'Cream Veranda' and 'Lavender Veranda'.

Introduced in 2007 by Kordes (KORfloci08).

Companions | The bright color of the blooms works well with other hot colors like reds, oranges, yellows, and golds. For red try 'Out of Rosenheim'; for yellow, 'Lemon Fizz' or 'Postillion'; for purple, 'Purple Rain' and 'Thrive! Lavender' complement the orange color. Some lighter pinks will set up the brilliant color of this rose.

Disease resistance	58
Flowering	25
Fragrance	0
Total rating	83

Large-flowered climber

Tall, upright to 7–8 feet

Salmon-pink with a yellow eye

I have grown and enjoyed this rose over the years for one simple reason—I don't have to do anything to it. 'Brite Eyes' just grows and blooms with little to no care. I have tried growing it on a fence, but prefer to grow it as a taller specimen plant in the border. The blooms are semi-double and open quite flat, revealing a wonderful brighter yellow color in the center, hence the name 'Brite Eyes'.

Introduced in 2006 by Will Radler (RADbrite).

Companions | 'Brite Eyes' can be grown as a large shrub or a small climber. The combination of a fairly dark pink with the yellow eye means that it can mix well with pinks, yellows, and whites. For rose companions try 'David Rockefeller's Golden Sparrow', 'Karl Ploberger', 'La Perla', 'Tupelo Honey', or 'Yellow Brick Road'.

'Brothers Grimm'

Disease resistance	60
Flowering	24
Fragrance	1
Total rating	85

Floribunda

Upright, bushy growth to 5 feet

Brilliant orange-red with gold reverses

Have you ever bought a plant because of the name? This one could be the first. The Brothers Grimm were 19th-century authors and academics who popularized such folklore tales as Cinderella, The Frog Prince, Hansel and Gretel, Rapunzel, and Snow White. This rose, much like the storytelling brothers, is a collection of desirable characters. The deep green, incredibly glossy foliage forms a backdrop for abundant, brilliantly colorful flowers, 3–4 inches across and borne in clusters. Their deep orange color fades to a dusty pink, and the petals have a wonderfully unexpected color reverse of golden-yellow. The flowers are fully double, almost old-fashioned in form.

'Brothers Grimm' is one of the healthiest, most vibrant roses I have had the pleasure of growing. It has won gold and silver medals at international rose competitions. I think it will be around for as long as the stories of the Brothers Grimm themselves.

Introduced by Kordes in 2002 as 'Gebrüder Grimm' (KORassenet).

Companions | The prominent growth and color of this rose means it could easily stand as a specimen in the landscape. Planted in mass, it is outstanding. To match some of its varied hues, try the roses 'David Rockefeller's Golden Sparrow', 'Golden Gate', 'Postillion', 'Purple Rain', and Tequila'.

Disease resistance	57
Flowering	20
Fragrance	7
Total rating	84

Hybrid Kordesii

Medium tall, with a strong spreading habit

Pink

This is a very hardy, fragrant variety that beautifully displays the strong characteristics of Kordesii breeding. The blooms are about 3 inches across, semi-double, and are usually produced in small clusters. Be sure to give this rose some space, as it really likes to grow wide and spread its wings. Another great introduction from the same breeder is 'Party Hardy'. Like 'Cape Diamond', it is bred for cold northern climates and is hardy to USDA zone 3.

Introduced in 2009 by Weeks Roses; hybridized by Christian Bédard.

Companions | I have used 'Cape Diamond' in the middle of the border, giving it enough room to develop its own shape and character. Other color blends that work really well with it include 'Cubana' and 'Peach Drift', as do roses in various shades of pink and white.

'Caramella'

Disease resistance 56
Flowering 22
Fragrance 2
Total rating 80

Shrub

Upright, bushy, and spreading to about 5 feet

Yellow-orange blend

'Caramella' was one of the first roses that I planted from the Kordes Fairy Tale series of roses. This series promised high disease resistance with beautiful, fully petaled, old-fashioned flowers. 'Caramella' has certainly fulfilled this promise. It is very vigorous in its growth with an upright presence in the garden that is not too overpowering. The color is a lovely, sunset-like blend of yellow to gold to amber and orange tints, hence the name 'Caramella' (caramel). The buds are a stunning deeper orange and open up with a high center, then unfurl with shades of pink in a beautiful, statuesque way, finally finishing with a yellowish-peach shade. Once the blooms open they show off the blousy, almost frilly petals to give a charming overall effect. The foliage is unique, extremely disease-resistant, with a greenish blue tint. This is truly a great rose that has won silver medals at the Kortrijk Rose Trials (2004) and the Hradec Rose Trials (2005).

Introduced in Europe in 2000 by Kordes; in the US in 2007 by Palatine Roses (KORkinteral).

Companions | The deep orange and pink buds let you have some fun finding saturated colors as complements. Good rose companions are 'Coral Drift', 'Flamenco Rosita', 'Mandarin Ice', 'Pomponella', and 'Summer Sun'.

Shrub

Upright, bushy growth to 4–6 feet

Medium pink

'Carefree Beauty' was a found rose known as 'Katy Road Pink'. It may have been found by a woman named Katy or it may have been found on a road named Katy—rose lore is sometimes unreliable. Gene sequencing revealed that it was one of Professor Buck's hybrids, created to survive the harsh winters on the Iowa prairie. Hardiness doesn't always imply toughness, but 'Carefree Beauty' is one tough rose. I have often given 'Carefree Beauty' as a gift because I know that the breeding of this rose will pave the road to success for the recipient. The bloom is somewhat soft in appearance, with only about fifteen to twenty petals that open up to a cupped, almost flat shape. These blooms seem to smile high above the disease-free foliage below. As a measure of its quality, this rose has been given the Earth-Kind designation.

Hybridized in 1977 by Griffith J. Buck (BUCbi).

Companions | 'Carefree Beauty' offers strength and flexibility in the landscape. It can serve well as a specimen in the mixed border or as a mass planting in a shrub border with excellent color. The pink of 'Carefree Beauty' can be paired with nearly anything and its casual blooms can work well with many other roses that have the same attribute, such as 'Lady Pamela Carol', 'Landmark Rose', 'Mother of Pearl', 'Mutabilis', or 'Old Baylor'.

Disease resistance	54
Flowering	23
Fragrance	1
Total rating	78

'Carefree Celebration'

Disease resistance	58
Flowering	26
Fragrance	2
Total rating	86

Shrub

Upright, bushy growth to 4–5 feet

Coral-orange

The Carefree series of shrub roses are medium-sized plants marketed by the Conard-Pyle company. The series is well named (although I doubt anyone would buy roses called 'Lots of Care'), as they are so easy to care for in the garden. Carefree roses are known for being disease-resistant and free blooming, often starting to flower in spring and continuing until frost. I have watched 'Carefree Celebration' grow vigorously and produce lots of color year after year, without having ever having to do much more than admire it. The color blend of orange and coral seems to be a favorite in the garden, always causing passersby to stop and look.

Introduced in 2007 by Conard-Pyle; hybridized by Will Radler (RADral).

Companions | I enjoy playing with the color of 'Carefree Celebration' by matching it with stronger red and orange tones, and oppositely with pastels. Try it with rose companions 'Cubana' (peach), 'Poseidon' (lavender), 'Postillion' (yellow), and 'Savannah' or 'Thrive! Lavender' (shades of lavender and pink).

'Carefree Delight'

Disease resistance	52
Flowering	25
Fragrance	0
Total rating	77

Shrub

Upright, bushy growth to 4–5 feet

Pink blend

The pink blend, single flowers of 'Carefree Delight' exude charm, and they are produced in large clusters that give a wonderful floral display on a continuous basis until frost. When autumn does arrive, this rose continues to delight with a nice display of hips. It has won medals at the Bagatelle Rose Trials and The Hague Rose Trials, and has been given awards by both the ADR and AARS.

Introduced in 1996 by Conard-Pyle; hybridized by Meilland (MEIpotal).

Companions | Match 'Carefree Delight' with other roses of the same floral size and landscape quality. I like to accentuate the pink color with other blends, and match the white center of this bloom with other whites. Good rose companions include 'Flamingo Kolorscape', 'Oso Happy Petit Pink', 'Peach Drift', 'Pink Drift', and 'Thérèse Bugnet'.

'Carefree Spirit'

Disease resistance	54
Flowering	25
Fragrance	0
Total rating	79

Shrub

Upright, bushy growth to 4–5 feet

Red with a white eye

This is often called the red 'Carefree Delight', but it is a different cross. However, with its cherry-red blend single flowers, this variety exudes much of the same charm as its relatives in the series. The dark flowers are produced in large clusters giving a continuous floral display until frost. The flowers open flat and reveal a white eye that shows off a vibrant display of stamens. The whole effect is charming. 'Carefree Spirit' won First Place at the 2007 Bagatelle Rose Trials and the Golden Rose Award at the Rose Hills International Rose Trials. It is also a 2009 AARS winner.

Introduced in 2009 by Conard-Pyle; hybridized by Meilland (MEIzmea).

Companions | The color of 'Carefree Spirit' stands out in the landscape more than other Carefree roses, and I like to match it with pinks or accentuate the white eye with other whites. Bring in some yellow to highlight the stamen display. Try 'Julia Child', 'Macy's Pride', 'Nastarana', 'Sea Foam', and 'Yellow Submarine'.

Disease resistance	56
Flowering	24
Fragrance	2
Total rating	82

Shrub

Bushy growth to about 3–4 feet

Light yellow

'Carefree Sunshine' is a great yellow addition to this series of otherwise pink or red roses. The soft yellow flowers are produced in clusters, opening flat to reveal a nice display of contrasting stamens and a blend of colors to a lighter yellow or creamy white.

Introduced by Conard-Pyle in 2001; hybridized by Will Radler (RADsun).

Companions | 'Carefree Sunshine' adds a swatch of brightness to any corner of the garden. I like to complement it with oranges, blends, and whites, such as the roses 'Garden Delight', 'Easter Basket', 'Innocencia Vigorosa', 'Mutabilis', and 'Summer Sun'.

'Carefree Wonder'

Disease resistance   50
Flowering   19
Fragrance   2
Total rating   71

Shrub

Upright, bushy growth to 4–5 feet

Pink

The pink of 'Carefree Wonder' is a medium tone blended with some white. The edges of the petals are somewhat pointed and add a lot of texture and dimension to the overall look. Like all the roses in this series, it is disease-resistant. It was given the AARS Award in 1991.

Introduced in 1991 by Conard-Pyle; hybridized in Meilland (MEIpitac).

Companions | I enjoy 'Carefree Wonder' with other landscape shrubs that have similar flower forms. I also like placing it with whites, pinks, and pink blends. Try it with 'Carefree Beauty', 'Easter Basket', 'Macy's Pride', 'Marie-Luise Marjan', and 'Old Baylor'.

Disease resistance	58
Flowering	23
Fragrance	1
Total rating	82

Grandiflora

Tall, bushy growth to 5 feet

Soft yellow to apricot

One of the Garden Art collection of roses, 'Centennial' has the desirable high-centered blooms of modern roses, which unfurl to reveal a wide palette of soft yellows to creamy white. These colors are set off beautifully by the plant's deep green foliage. It's a good rose in a wide range of climates, especially in northern gardens.

Introduced in 2006 by Bailey Nurseries; hybridized in 1996 by Ping Lim (BAIcent).

Companions | The soft shades of 'Centennial' allows for combinations with a multitude of colors. Try it with some hot tones such 'All the Rage', 'Bajazzo', 'Brothers Grimm', 'Oso Easy Cherry Pie', and 'Summer Sun'. Or keep the color palette in soft tones with whites and pastels. Basically, you can't go wrong with this rose.

'Cinderella'

Disease resistance   60
Flowering   20
Fragrance   9
Total rating   89

Shrub

Tall, upright growth to 6 feet or more

Pink

For a rose to be called 'Cinderella' it had better be lovely, and this one is definitely lovely. One of the Fairy Tale series of roses, 'Cinderella' is a pretty pink with clusters of wonderfully frilly, full, old-fashioned flowers that cover the plant like a gown worn to the prince's ball. And as befits a would-be princess, the fragrance is perfect. This is fantastic grower that can get quite tall, with deep green, extremely healthy foliage. In my experience of growing this rose I can't remember having seen any black spot on it.

Introduced in 2006 by Kordes-Newflora (KORfobalt).

Companions | I've grown 'Cinderella' in a bed and in the back of the border near a fence where it can grow and develop into a big, beautiful ball gown of a rose. You can also plant it where the fragrance can be enjoyed. The soft pink combines beautifully with any pastels. The pink fades to a lighter shade that is almost white, so you can also put this rose with other whites like 'KOSMOS', 'Lion's Rose', and 'Summer Memories'. Other pinks like 'Alexandra Princesse de Luxembourg' and 'Eliza' are sufficiently tall and upright to hold their own near 'Cinderella'.

Disease resistance	53
Flowering	23
Fragrance	5
Total rating	81

Climber

Climbing to 10–15 feet

Pink

'Climbing Pinkie' has been in the trade for over half a century and it has more than proven itself as a worthy garden rose, so much so that it has been given the Earth-Kind designation. Personally, I have enjoyed this rose in the garden simply because of its flexibility. The canes are so pliable that it is easy to wrap them around a pillar or arbor. On a fence it can easily be trained in any direction. An added bonus is that 'Climbing Pinkie' has few thorns and is fragrant, so it is a good choice for an arbor or archway where there is plenty of foot traffic.

The blooms themselves are not very complex but they are produced in huge clusters and it is the whole here that is greater than the sum of the parts. All of these blooms together on a support are beautiful and they add charm in a way that seems like the rose has been there for a lifetime.

Discovered in 1952 as a climbing sport of a popular Polyantha, 'Pinkie', by E. P. Dering.

Companions | Grow 'Climbing Pinkie' on a fence, post, arbor, or pergola—it all works for this rose. You can accompany this climber with other pinks, pastels, and whites. Try it with 'Blush Noisette', 'Cubana', 'Peach Drift', 'Savannah', and 'Sweet Fragrance'.

'Coral Drift'

Disease resistance	60
Flowering	26
Fragrance	0
Total rating	86

Shrub

Short and spreading habit to 2 feet

Coral-orange

The Drift roses are a cross between full-size groundcover roses and miniatures. From the former they inherit toughness, disease resistance, and winter hardiness. From their miniature parentage, they inherit their well-managed size and repeat-blooming nature. You can use Drift roses in the garden in massed plantings, in groups of three or five, in a pot, or along a path near the front of the border.

'Coral Drift' is one of the most consistent roses in the garden. It is consistent in bloom production; in fact, it is nearly always blooming. It is consistent in its ability to resist disease; I have never seen any black spot on this variety. In rose trials 'Coral Drift' receives superior scores year after year. The blooms are about 1 inch in diameter. They open up in a softly cupped form to reveal the stamens.

Their bright orange color really pops above the deep green, glossy foliage.

Introduced in 2008 by Conard-Pyle; hybridized by Meilland (MEldrifora).

Companions | Companion plants with lavenders and blues work well with the orange shade of 'Coral Drift', so try it with 'Thrive! Lavender'. The bright color also shows well with other hotter colors and whites, including 'Garden Delight', 'Lion's Rose', and 'Michel Bras', or the softer tones of 'Caramella'.

Disease resistance	49
Flowering	23
Fragrance	4
Total rating	86

Floribunda

Shorter, bushy, and spreading to 2 feet

Creamy pastel pink

The flowers of this plant are charming and old-fashioned, with a subtle pastel coloring of cream into softest pink. The flower size is pleasingly large for the size of the plant. They are multi-petaled and sometimes present with a button eye. I have witnessed occasional black spot on this rose (depending on weather and temperatures) but I always forgive the plant its spots when I see the blooms. They are completely charming in color and effect. 'Cream Veranda' is part of the Veranda series of roses, all very good roses for the patio, pots, or front of the border. It was given an ADR Rose Award in 2009, and has won other international competitions including a silver medal at Le Roeulx in 2008.

Introduced in 2007 by Kordes (KORfloci01); also sold as 'Garden of Roses'.

Companions | The soft pastel color of this rose offers a multitude of possibilities for color combinations. I have found it blends beautifully with reds, oranges, apricots, and pinks. The old-fashioned blooms of 'Out of Rosenheim' will keep that nostalgic feeling along with 'Caramella' and 'Cinderella'. 'Larissa' will offer a different flower size. 'Lion's Rose' will pick up on the subtle peach coloring.

'Crimson Meidiland'

Disease resistance	**60**
Flowering	**26**
Fragrance	**0**
Total rating	**76**

Shrub

Bushy growth, spreading to 2–3 feet

Crimson

The Meidiland Landscape series of roses are truly wonderful for the chemical-free garden. They rarely suffer from disease, they have wonderful glossy foliage, and the flowers keep coming over a long season. The individual flowers are small but they are produced in large numbers on a panicle of bloom. These are great roses to plant en masse, as they provide a wonderful texture and color complement to many other roses. Their spreading habit allows them to weave together and help to suppress any weeds. Other great roses in this series include 'Fairy Meidiland', which is a lovely non-fading pink; bright red 'Fire Meidiland'; 'Ruby Meidiland' with double ruby red clusters; and 'White Meidiland', which has small white blooms that look like a blanket of snow.

Introduced in 2007 by Conard-Pyle; hybridized by Meilland (MEIzerbil).

Companions | For the best effect, plant the Meidiland roses in numbers or mass along borders and pathways for the greatest effect as their textures and colors will be a foreground complement to many other roses. For long-lasting color, plant in containers. Use 'Crimson Meidiland' to add texture to plants with flowers that include other reds, dark colors, blends, or even as a contrast to pastels and whites. Good companion roses include 'Caramella', 'Grande Amore', 'Mother of Pearl', 'Thanksgiving Rose', and the Thrive! and Knock Out roses.

Shrub

Compact, bushy, and spreading to 3 feet

Apricot blend

I really love roses like 'Cubana' because they look terrific on their own but they also pair well with other varieties with a different shape or texture (once more, the whole is greater than the sum of the parts). 'Cubana' has an open flower form with petals that have a ruffled look and a generous presentation of contrasting stamens. Because this rose produces so many flowers, its overall effect is like a frilly carpet of apricot fading to pink. The foliage is generally dark green and glossy, but I have seen 'Cubana' get black spot under conditions that favor the fungus. I recommend planting it near another resistant variety that is complementary in color, to reduce the local disease pressure. This is a rose that is not grown often enough, but it should be.

Introduced in 2007 by Palatine-Kordes (KORpatetof).

Companions | The growth habit of 'Cubana' suggests a front-of-the-border position complemented by another variety behind it. Try 'Mandarin Ice', 'Mother of Pearl', 'Summer Sun', 'Thérèse Bugnet', and 'Zaide'.

Disease resistance	**49**
Flowering	**25**
Fragrance	**0**
Total rating	**74**

'Dark Desire'

Disease resistance	**55**
Flowering	**28**
Fragrance	**10**
Total rating	**93**

Hybrid Tea

Bushy, upright habit to 4 feet

Violet-red

More than twenty years ago, the hybridizers at the Kordes rose company decided that breeding healthy, no-spray roses should be their primary ambition, and that is still the goal with every rose they grow today. 'Dark Desire' is part of their Parfuma collection and it is a beautiful, sensuous rose—everything you could want from a Hybrid Tea. From buds that are almost black, the cupped flowers emerge with a unique violet-red color before opening into a cupped bloom that displays a spectrum of color from red to violet to purple to even darker hues. The foliage is very resistant to fungal diseases and is a bright glossy green. The fragrance of this rose can be described as sophisticated, with intertwining floral and fruity aspects. With its classic form, the deep rich color, and sensual fragrance, this rose is indeed desirable. Other roses in the Parfuma collection also belie the contention that modern roses have no fragrance; 'First Crush' and 'Summer Romance' are examples.

Introduced in 2014 by Kordes (KORdiagraf); also sold as 'Grafin Diana'.

Companions | 'Dark Desire' fits well in the middle to the front of the border where its fragrance can be appreciated. The deep colors offer a lot of flexibility when it comes to blending with other tones. You can stay with solid, saturated colors by pairing it with roses such as 'Eifelzauber' and 'Poseidon'. Other roses that carry a similar violet-red color in their flowers or buds are 'Caramella', 'KOSMOS', and 'Raspberry Kiss'.

Disease resistance	60
Flowering	23
Fragrance	7
Total rating	90

Hybrid Musk

Tall, bushy growth to 7–10 feet

White

This is one of my all-time favorite roses. Traditionally, when someone discovers a rose that can't be identified, they have the right to name the rose. ‘Darlow's Enigma’ was discovered circa 1995 by Mike Darlow, and he clearly couldn't discern much about its provenance, hence the name. No matter how this rose is identified, I am just thankful that it is around for gardeners to enjoy.

The reason I love this rose so much is its generosity in the landscape.

I have grown ‘Darlow's Enigma’ in a multitude of gardens and it just keeps on giving. It grows quite tall and bushy, with deep green foliage that is always healthy and vibrant. I don't think I have ever seen any black spot on this rose. The flowers are small, consisting of seven to eight petals that are arranged in the most charming way. They open flat to perfectly frame a burst of yellow stamens that seem to have so much presence and energy. The effect is magnified as the flowers are produced in clusters that follow one after another. The fragrance can be described as strong.

Companions | ‘Darlow's Enigma’ is tall, so give it some room in the back of the border or as a specimen plant in the landscape. It has quite stiff canes so it is not a great candidate for wrapping around a post or tying to a fence. The simplicity of its flower and the pure white color will steer you in the direction of other flower forms and colors. For softer colors, try ‘Blush Noisette’ and ‘Wedding Bells’. For brighter colors try ‘Oso Easy Cherry Pie’, ‘Ruby Ice’, and ‘Thérèse Bugnet’.

'David Rockefeller's Golden Sparrow'

Disease resistance	55
Flowering	20
Fragrance	0
Total rating	75

Shrub

Tall, with an arching, spreading habit to 4 feet wide and tall

Butter yellow

One of the great joys of my position at the NYGB was the opportunity to spend some time with David Rockefeller. When I first met him, we had just completed the 2007 renovation of the Peggy Rockefeller Rose Garden and I was thrilled when he told me that his late wife would have loved the rose garden. It was a pleasure to work with the breeder to have this rose named in honor of Mr. Rockefeller. 'Goldspatz' means golden sparrow in German, and because yellow is Mr. Rockefeller's favorite color, we renamed it 'David Rockefeller's Golden Sparrow'.

The rose itself has proven to be extremely healthy with glossy, deep green foliage on long arching canes. The soft yellow blooms are produced in clusters starting with a rich butter yellow with matching stamens and then fading to a lighter yellow as they age. This rose has an understated charm and gentleness about it that is hard to describe, much like my experience with Mr. Rockefeller himself.

Introduced in 2011 by Kordes-Newflora as 'Goldspatz'; renamed 'David Rockefeller's Golden Sparrow' in the US (KORgellan).

Companions | I have grown this rose in large containers where it can act as a focal point with its arching canes gently bobbing in the wind. The growth habit also allows for a lot of flexibility in the mixed border. Complementary pastels are wonderful with this rose; I like it with the soft pink of 'Larissa', or with whites like 'Escimo', 'Innocencia Vigorosa', and 'Summer Memories'. A hotter color to consider is the deeper peach of 'Summer Sun'.

'Dee-Lish'

Disease resistance	49
Flowering	13
Fragrance	5
Total rating	67

Hybrid Tea

Tall, upright growth to 6 feet

Pink

As is the case with many Hybrid Teas, 'Dee-Lish' scores a bit low due to the simple fact that it receives lower scores for blooming than a similar shrub rose. But this rose is an ADR winner, and it offers lovely color on a classic form—with fragrance to boot.

The Conard-Pyle catalog describes the fragrance of 'Dee-Lish' as verbena and citrus. The deep pink blooms open to a very old-fashioned type of flower, which is charming with its tints of silver. Many of my rose-loving friends in the South rave about 'Dee-Lish', which makes me think it is even more delicious and disease-resistant in hotter climates.

Introduced in 2014 Conard-Pyle; hybridized by Meilland (MEIcludif).

Companions | 'Dee-Lish' is a tall grower, so put it in the back of the border in a place where you can still enjoy the fragrance. Match it with lavender-colored roses such as 'Alexandra Princesse of Luxembourg', 'Poseidon', and 'Wedding Bells', or complement it with a nice pink like 'Larissa'.

'Dolomiti'

Disease resistance	58
Flowering	22
Fragrance	0
Total rating	80

Floribunda

Bushy growth to 3 feet

Pink with a creamy white eye

Although it is not seen as often in North American gardens as it is in European ones, this rose has been a joy to grow. I love the simple, open flower form and the wonderful color contrast between the pink, ruffled edges and the creamy white eye. As an added bonus, the white eye is a fantastic background to display the long stamens like a burst of fireworks. The whole effect is a wonderful play of colors. The foliage is dark green and glossy, and I have not seen any black spot on this variety so far. In fall, 'Dolomiti' is a good variety for rose hips.

'Dolomiti' has won many awards both nationally and internationally, including an ADR Rose Award in 2009, and gold medals in Baden-Baden (2008 and 2009) and at the Kortrijk Rose Trials (2013).

Introduced in 2011 by Kordes (KORrahibe).

Companions | The compact habit of 'Dolomiti' makes it a good choice for a pot or in the front of the border mixed with other pinks, oranges, and whites. For pinks try 'Belinda's Dream', 'Care-free Beauty', and 'Thérèse Bugnet'. For blending with other orange tones, plant it near 'Coral Drift' or 'Michel Bras'.

Disease resistance	56
Flowering	21
Fragrance	5
Total rating	**82**

China

Bushy growth to 3–5 feet

White

Like most China roses, 'Ducher' does best in warmer climates. Although I have successfully grown China roses in zone 6, these roses do need a warm spot, perhaps in a sunny, sheltered microclimate in your yard. I include 'Ducher' in this book along with 'Mutabilis', another China rose, for their proven durability in the landscape. Like most China roses, 'Ducher' is constantly pushing out new growth and flowers in what seems to be an unending array of nooks and crannies in the plant. The white blooms open up to a more open form but never really seem to reveal their stamens. Perhaps in the hottest situations the blooms would open more fully. Even without this, though, the constant supply of pure white blooms is refreshing and lightens up any corner of the garden. The fragrance of 'Ducher' is typically described as soft and delicate. For its toughness and disease resistance, 'Ducher' has been given the Earth-Kind designation.

Hybridized in 1869 by Jean-Claude Ducher.

Companions | 'Ducher' is a versatile rose. It makes a great specimen plant in the mixed border, it can form a good hedge, or it can be in the spotlight in a container. Plant it near the border edge where you can enjoy the fragrance. In some cases there will be a little pink flush to the buds, so play off this by planting it with other pale roses like 'Blush Noisette', 'Ole', 'Pink Drift', and 'Pink Pet'. Or plant it with 'Mandarin Ice' or 'Ruby Ice' to pick up on the white reverse of their petals.

'Easter Basket'

Disease resistance	**51**
Flowering	**25**
Fragrance	**0**
Total rating	**76**

Floribunda

Shorter, with a bushy habit to about 3–4 feet

Creamy white to yellow

To coin a popular phrase, "This rose had me at hello!" The colors of the blooms are such an enchanting blend of pink, yellow, and cream that the name 'Easter Basket' is a perfect description. Plus this rose has what people call flower power—it is rarely out of bloom. The individual blooms are a creamy white with a slightly yellow tint in the center. As they open and age, the petals gain a pink, ruffled edge that adds texture and lends the plant a somewhat nostalgic look. Like most Floribundas, the blooms are produced in clusters of three to five per stem. All of this floral display is atop nicely green foliage.

Although I have seen some black spot on this variety during very wet and stormy seasons, 'Easter Basket' consistently scores high in trials for disease resistance. The constant floral display of this rose is enough to overlook the occasional black spot.

Introduced in 2007 by Meilland (MEIpoten).

Companions | 'Easter Basket' can be a flowering powerhouse, so plant it with other complementary colors and enjoy the Easter parade! Planted as a hedge or in groups, it can make quite a presence. In the border, place it near the front to middle. Play off the pink and yellow colors by putting it with 'Belinda's Dream', 'Eliza', 'Jane Bullock', 'Pink Drift', and 'Thrive! Lemon'.

Floribunda

Bushy habit to about 3–4 feet

Orange and pink blend

Easy does it? Yes it does. This variety truly requires little care. The multi-colored blooms are a mixture of hues that you might see in a never-ending sunset. The petals unfurl to reveal a frilly, scalloped edge, which seem to add warmth and texture to the bloom as they unfurl. Plant it in a pot where you can enjoy the color blend close up, or plant it in a group to enjoy day-long sunset in your garden. 'Easy Does It' is a 2010 AARS award winner and has received a number of international awards, including Best in Trial at the Australian National Rose Trials and a People's Choice Award at the Rose Hills International Rose Trials.

Introduced in 2010 by Weeks Roses–Bailey Nurseries (HARpageant).

Companions | Try some complementary colors with roses such as 'Bajazzo', 'Beverly', 'Black Forest Rose', 'Brothers Grimm', 'Fiji', and 'Flamenco Rosita'.

Disease resistance	**52**
Flowering	**26**
Fragrance	**5**
Total rating	**83**

'Eifelzauber'

Disease resistance	53
Flowering	17
Fragrance	2
Total rating	72

Shrub

Tall, upright, and bushy, reaching to 5 feet

Lighter pink outer petals fading to creamy white

This rose has taken a little while to establish in my garden but now that it has, I am so glad it is there. First of all, the pastel color of this pink (which could even be described as soft apricot) is so pleasing. The color fades with age, leaving a whole spectrum of darker to lighter flowers on the plant all at once. The flowers are borne in large clusters atop of green, glossy foliage, and the flower form throws in some nostalgia and frilliness. 'Eifelzauber' does well in a range of climates and it deserves to be more widely planted.

Introduced in 2008 by Kordes (KORcarbas).

Companions | The softness of this pinky white allows you to take your planting scheme in all kinds of directions. Try 'Eifelzauber' with reds such as 'Black Forest Rose'. To play up the old-fashioned look, try 'Alexandra Princesse de Luxembourg' or 'Eliza'. 'Dolomiti' and 'Larissa' nicely complement the flower form. A combination with the soft lavender of 'Poseidon' would be a showstopper.

Floribunda

Upright and bushy to 4 feet

Lightest pink to creamy white

This rose is in the Fairy Tale series—roses that give a nod to the nostalgic look of old roses, but with repeat blooming and excellent disease resistance. A unique characteristic of this rose's flowers is how much they resemble camellia blooms when opening. In my garden, 'Elegant Fairy Tale' has taken a little while to get established, but it has been worth the wait. In warmer climates, it can get quite large.

Introduced in 2000 by Kordes (KORterschi); also sold as 'Bremer Stadtmusikanten', 'Belami', and 'Pearl'.

Companions | There are so many possibilities of color combinations with this rose, from white and pink to orange, purple, and red. I have particularly enjoyed pairing it with some darker pinks like 'Eliza'. 'Pink Bassino', 'Pomponella', and 'Fortuna Vigorosa' provide a complement with their contrasting flower forms. The softer pink of 'Quietness' perfectly pairs with the lighter tones of 'Elegant Fairy Tale'.

Disease resistance	50
Flowering	17
Fragrance	3
Total rating	**70**

'Eliza'

Disease resistance	59
Flowering	24
Fragrance	2
Total rating	85

Hybrid Tea

Tall, upright growth to 6 feet

Silvery pink

I have often suggested 'Eliza' as a substitute for gardeners who love 'Queen Elizabeth' but find it to be disease-prone. 'Eliza' grows quite tall and vigorous. It is classified as a Hybrid Tea, but the blooms are often borne in clusters. The flowers are a deep, rich pink and show many gradients of color all the way to silver so the whole effect is more of a pink blend. The flower petals themselves often have a scalloped edge. The foliage is incredibly healthy; in fact, I don't recall ever seeing any black spot on this rose. The deep green, glossiness of the leaves is a perfect background for the large blossoms, and the new foliage growth has a lovely reddish hue.

A must-have rose for any garden, 'Eliza' has won many awards, including an ADR Rose Award (2005), a gold medal and Prize of the City in Belfast (2006), a gold medal in Dublin, and other awards at trials and competitions at The Hague, Hradec, St. Albans, and Toproos.

Introduced in 2004 by Kordes (KORaburg).

Companions | Because 'Eliza' can grow quite tall, use it as a specimen plant in the middle to back of the border. The base pink and gradients of pink in the bloom allow for great combinations with other roses and flowering plants. Any white or lighter pinks would highlight this color scheme, including 'Alexandra Princesse de Luxembourg', 'Cinderella', 'Easter Basket', 'Lion's Rose', and 'Macy's Pride'.

Shrub

Bushy and spreading to about 3–4 feet

White

This is a hardy and a vigorous grower that sends up strong canes topped with a multitude of buds that are held high above the foliage. 'Escimo' bears pure white, single flowers that simply and stunningly show off yellow stamens in the center of each bloom. With the flowers opening flat, in full bloom it looks as though the plant is covered with a blanket of snow. The foliage is deep green, very glossy, and healthy. A few times I have seen some leaf spot on this plant, but never anything that the plant can't survive.

'Escimo' is much-heralded in rose trials and has won many awards, including the ADR Rose Award (2006), the Australian National Rose Trials silver medal (2011), the Royal National Rose Society President's International Trophy (2001), and gold and silver medals in other international rose competitions.

Introduced in 2006 by Kordes (KORmifari).

Companions | The growth habit of 'Escimo' suggests placement near the front of the border, alongside a pathway, or in a container. Planted in mass, it creates a stunning white rose hedge that is rarely without flowers. Because of the pure white color, virtually anything can color-match with 'Escimo'. I have grown it with other yellows to play off the wonderful stamen display. Try 'Centennial', 'Golden Fairy Tale', 'Postillion', and 'Winter Sun'.

Disease resistance	52
Flowering	26
Fragrance	0
Total rating	78

'Felicitas'

Disease resistance	56
Flowering	27
Fragrance	0
Total rating	83

Shrub

Medium height and spread to about 3–4 feet

Pink

In terms of care, 'Felicitas' is one of those plant-it-and-forget-it roses. In terms of style, you can't forget it because it keeps growing and producing these healthy, long canes and a profusion of simple pink flowers in huge clusters. How can you not take notice of such giving in a rose? The color is a dusty pink and the flower is a simple form, with five petals revealing a gorgeous posy of stamens. The flowers seem to smile, and I have noticed that the bees are crazy for them. The overall effect of this rose is a charming swath of pink in your garden. The growing habit of 'Felicitas' lends itself to mass planting as a dense, colorful shrub or hedge. A few plants will spread and mesh together forming a tight clump of glossy, disease-free foliage and hundreds of pink flowers. 'Felicitas' won the ADR Rose Award in 1996.

Introduced in 1998 by Kordes (KORberis).

Companions | Plant 'Felicitas' as a hedge or along a walkway or border. Its spreading habit means that it needs some room. The pleasing pink color is easy to blend with nearly anything. One idea is to try highlighting the impressive orange stamens with other oranges or peach blends. Try for this effect by pairing it with 'Caramella', 'Coral Drift', 'Cubana', 'Mother of Pearl', and 'Summer Sun'.

Disease resistance	**60**
Flowering	**25**
Fragrance	**5**
Total rating	**90**

Hybrid Tea

Bushy, dense growth habit to about 3–4 feet

Cherry pink

'Fiji' is at the front of the line of new disease-resistant Hybrid Teas, and like its island namesake, this rose will captivate you much like a perfect holiday with tropical islands and sandy beaches. The brilliant cherry pink color of this rose certainly makes me think of a fruity delicious drink.

The scalloped edges on its petals add unexpected welcome and texture, and although 'Fiji' is classed as a Hybrid Tea, these unusually shaped flowers appear in small clusters that give the whole effect a much greater presence in the garden. This rose has been featured in numerous international rose competitions where it received silver medals in Mona and Kortrijk (2012), and further awards at the Bagatelle, Le Roeulx, and Lyon rose trials.

Introduced in 2012 by Kordes (KORladcher); also listed as 'Cherry Lady'.

Companions | Plant this variety close to the front of the border where you can enjoy the charm of the blooms and some added fragrance. Complement the tropical colors with other blends, like 'Caramella', 'Easter Basket', 'Easy Does It', and 'Postillion'. Any white will do as well.

'First Crush'

Floribunda

Bushy growth to 3 feet

Light, creamy pink

Not only does this rose have high disease resistance, but also 'First Crush' satisfies the senses in every way. It produces clusters of lushly fragrant blooms. As the buds begin to open they reveal a dark red, almost plum color, but the petals quickly unfurl to a most delicate creamy pink. The combination of buds and open blooms in a single cluster is really beautiful. To my eye, the whole effect is a throwback to the roses of bygone days. With the German name 'Constanze Mozart', I imagine this rose growing in Mozart's own garden, while the sound of his sonatas echoes through the garden paths.

Introduced in 2012 by Kordes as 'Constanze Mozart' (KORmaccap).

Companions | This rose is a perfect candidate for a container where its flowers can be raised so that the fragrance can be enjoyed up close. In the garden, the growth habit means that it fits nicely in the middle to front of the border. The color is flexible yet delicate so be careful not to overwhelm it with a surrounding palette of stronger tones. Try some muted reds to pick up on the bud color, such as 'Miracle on the Hudson'. Other roses with soft colors include 'Alexandra Princesse de Luxembourg', 'Eifelzauber', 'Marie-Luise Marjan', and 'Poseidon'.

Disease resistance	50
Flowering	25
Fragrance	10
Total rating	85

Disease resistance	49
Flowering	23
Fragrance	3
Total rating	75

Shrub

Bushy and upright to about 4 feet

Peachy yellow

One of the things I most enjoy about 'F. J. Lindheimer' is the color play of the blooms. Let me preface that statement by saying this variety is virtually never without blooms, producing a constant display that gives the overall effect of many colors at once, from yellow to peach to soft orange. The foliage is of consistent health and color and provides a steady background for the flower display.

Introduced in 2004 by the Antique Rose Emporium.

Companions | This variety is so gracious in bloom production and color you can try accenting the constantly changing palette of yellows, peaches, and soft oranges with similar hues. For orange and peach, pair it with 'All the Rage', 'Garden Delight', or 'Michael Bras'. For a yellow blend, try 'Centennial' and 'Lady Pamela Carol'.

'Flamenco Rosita'

Disease resistance	57
Flowering	27
Fragrance	2
Total rating	86

Shrub

Bushy growth to 4–5 feet

Dark pink to cherry-red

'Flamenco Rosita' came to my attention when some rose-growing friends in Texas told me that it was definitely a variety to try. I'm grateful for the recommendation because 'Flamenco Rosita' has been impressive indeed. Its growth is upright, vigorous, and healthy. The flower production is also impressive—there always seem to be these beautiful, full, dark pink to red blooms to enjoy. The individual blooms open flatter to a heavily petaled, old-fashioned look. The blooms are produced in clusters and seem to be never-ending, perhaps portraying the clapping and music of the flamenco dance that never wants to stop.

Hybridized in 2006 by Amanda Beales (BEAdonald).

Companions | With "flamenco" in its name, you can expect hot colors of pink to red, so try this variety with other pinks, blends, and whites, including 'Caramella', 'Mutabilis', 'Raspberry Kiss', and 'Rosenstadt Freising'. For an even bolder combination, put it with orange-peach 'Garden Delight'.

Disease resistance	50
Flowering	26
Fragrance	0
Total rating	76

Shrub

Short and bushy to about 3 feet

Medium pink

As with Will Radler's Knock Out collection of roses, the Kolorscape collection by Kordes will inspire confidence even in the inexperienced gardener. These roses are compact, mounded shrubs that give a consistency of color from first bloom until frost. The foliage is glossy and green. They have a generous bloom cycle that will continue through the season. Kolorscape roses have self-cleaning characteristics, which means their petals fall cleanly from the plant after the blooms are done. They have been tested for heat tolerance throughout the United States.

'Flamingo Kolorscape' has a semi-double flower form with about ten to fifteen petals, which open flat to reveal the stamens inside. With its bright pink color, it provides a wonderful splash of color in the garden. This rose was awarded an ADR Rose Award in 2013.

Introduced in 2012 by Kordes (KORhopiko).

Companions | All the roses in the Kolorscape series are versatile. They work well planted in the front of the border, along a pathway, or in a mass planting, or as a rose hedge. In the garden, this rose works well with other roses that have pastel colors, deep pinks, or whites. Try it with 'Easter Basket', 'Elegant Fairy Tale', 'Garden Delight', 'Marie-Luise Marjan', and 'Mutabilis'.

'Flirt 2011'

Disease resistance	55
Flowering	25
Fragrance	0
Total rating	80

Miniature

Short, compact growth to 2 feet

Pink with white tones

This 2011 ADR award winner is a joyful addition to northern gardens. The name 'Flirt 2011' is appropriate, as the color play does seem almost flirtatious and it is extremely floriferous, constantly producing charming little flowers from first flush until frost. The foliage is a wonderful steady green, very healthy and resistant to diseases, which allows for a nice canvas for the flowers. The individual flowers themselves are small, borne in clusters, so the overall effect is a lot of color for the size of the plant. The flowers open to a captivating cup shape and reveal a creamy white center and base of the petals. The petals themselves have the same creamy white reverse and they tend to have a little scallop or frill about them, lending extra texture and depth. They almost resemble little pink-and-white frilly umbrellas. This rose is fun and flirty all the way around.

'Flirt 2011' has won medals at the Buenos Aires Rose Trial (Gold, 2010), a Certificate of Merit at the Royal National Rose Society, and the Belfast Rose Trials Best Miniature or Patio Rose (2013).

Introduced in 2011 by Kordes (KORchakon).

Companions | The compact size of 'Flirt 2011' allows it to be enjoyed in the front of the border, in a mass along a pathway, or in a container. The color scheme can be matched with other pinks, whites, and pastel blends. Try it with 'Ducher', 'Old Blush', 'Peach Drift', 'Pink Drift', and 'Stanwell Perpetual'.

Large-flowered climber

Climbing to 8–10 feet

Red

It is great to have climbing roses in this color with good disease resistance, because that has not always been the case. 'Florentina' has very healthy and hardy foliage that forms a beautiful backdrop to the nostalgic novelty of its flowers. These are beautifully cupped, with a wonderful petal display all surrounding the yellow stamens like a perfectly wrapped present. The petals themselves have a ruffled effect, which adds appealing depth. Planted on a light post or pillar, 'Florentina' is a feast for the eyes. It was awarded the Buenos Aires Rose Trials gold medal in 2012.

Introduced in 2011 by Kordes (KORtrameilo).

Companions | The growth habit and bold color of 'Florentina' means that it makes a strong statement on a fence, pillar, or arbor. The deep, rich red is compatible with a lot of color combinations. Try some other saturated colors or red blends, such as 'Mutabilis', 'Oso Easy Cherry Pie', 'Ruby Ice', and 'Thanksgiving Rose'.

Disease resistance	50
Flowering	24
Fragrance	0
Total rating	74

'Flower Carpet Amber'

Disease resistance	**58**
Flowering	**22**
Fragrance	**0**
Total rating	**80**

Shrub

Short and bushy to 2–3 feet

Apricot

The Flower Carpet series of roses has been around for some time. These low-growing roses have a long flowering period, are easy to care for, and have exceptional disease resistance. "Carpet" implies that these are ground-hugging or spreading plants, but this is generally not the case. However, when planted in a mass alongside a drive or boulevard their growth habit does cover the ground in such as way that they do look resemble a carpet of color that is about 2 feet high.

A series of roses sometimes appears to offer the same rose but in many different colors. However, it has been my experience that roses of certain colors are less resistant to disease than others, particularly yellow. As a result, I have found that although most of the Flower Carpet roses have good disease resistance, 'Flower Carpet Yellow' is not one of them. The later introductions (of which 'Flower Carpet Amber' is one) seem to be the best performers in terms of disease resistance. Other good roses in the series include 'Flower Carpet Pink Supreme' and 'Flower Carpet Scarlet'.

'Flower Carpet Amber' is a charming rose with shades of apricot and pink with a bright center of yellow stamens. The whole effect is subtle, yet colorful and cheerful. It was given the ADR Rose Award in 2009.

Introduced in 2005 by Noack.

Companions | The smaller growth habit means that 'Flower Carpet Amber' is a good variety along a pathway, a drive, in the front of the border, or in a container—but it is when planted in numbers that these roses really shine in the landscape. The color of this rose is fun to play with; try other blends, pinks, and yellows like 'Eifelzauber', 'Garden of Roses', 'Savannah', 'Sunny Sky', and 'Sweet Jane'.

Disease resistance	56
Flowering	27
Fragrance	0
Total rating	83

Floribunda

Short and spreading to 2–3 feet

Salmon-pink

This is another outstanding variety in the Vigorosa series of roses from Kordes. Once again, it's a simple flower form that can charm anybody with its simple beauty. When this plant is in full bloom the foliage is completely obscured by a mass of subtle pinks. 'Fortuna Vigorosa' is fantastic in mass or as a single plant near the front of the border or along the walkway.

Introduced in 2002 by Kordes (KORatomi).

Companions | This is a great rose for mass plantings or spaces along pathways or in the front of the border. This color will mix with a variety of other tones, but it truly shines when paired with a more muted palette. The lighter eye in the center of the bloom goes beautifully with cream and white roses like 'KOSMOS' and 'Lion's Rose'. Or try echoing the yellow stamens by planting it in front of roses with similar colors such as 'Karl Ploberger', 'Sunny Sky', and 'Winter Sun'.

'Francis Meilland'

Hybrid Tea

Tall, upright growth to 6 feet

Shell pink fading to white

Created by Michou Meilland, the daughter of the rose breeder Francis Meilland, this is a fitting tribute to the creator of the legendary 'Peace' rose. 'Francis Meilland' is a perfect combination of health, vigor, flower form, color and fragrance, which is why it has won many awards, including an ADR Rose Award in 2008, a gold medal at the Monaco Rose Trials in 2011, and an AARS award in 2013. It has blooms that sit high above very strong, disease-free foliage. Plant this rose where you can immerse yourself in the delightful fragrance.

Introduced in 2013 by Meilland (MEItroni); also sold as 'Pretty Woman', 'Prince Jardinier', and 'Schloss Ippenburg'.

Companions | Be aware that this variety can grow tall, so keep it toward the back of the border. Try it with other pastels and whites, like 'Cinderella', Easter Basket', 'Eliza', 'First Crush', 'KOSMOS', 'Lion's Rose', and 'Summer Memories'.

Disease resistance	50
Flowering	20
Fragrance	8
Total rating	78

Disease resistance	58
Flowering	27
Fragrance	1
Total rating	86

Floribunda

Tall and bushy to 5 feet

Yellow-gold with orange-pink edges

Is this the color of the sunrise or the sunset—or both? 'Garden Delight' is probably one of the most photographed roses in my garden, stopping visitors in their tracks with sheer enjoyment. It displays a multitude of bright colors all at once: the bud opens golden-yellowy orange with edges of darker oranges and pinks. The flowers are many-petaled and bloom in large sprays. This plant itself grows vigorously and I have seen very little problem with fungal diseases. I have also seen this rose sold under the names 'Garden Fun' and 'Garden Joy', both of which suit it perfectly.

Introduced in 2009 by Kordes (KORgohowa); also called 'Gartenspass'.

Companions | The strong colors of this rose lend it to planting with other bright colors. Keep the sunset in mind and try reds such as 'Black Forest Rose', or oranges and yellows like 'David Rockefeller's Golden Sparrow', 'Mandarin Ice', and 'Sunny Sky'. Because the finish color on the edges is pink, believe it or not pinks work well. Try it with 'Carefree Beauty' and you'll see what I mean!

'Garden Sun'

Disease resistance   51
Flowering   22
Fragrance   3
Total rating   76

Large-flowered climber

Climbing to 10–12 feet

Apricot-peach blend

Good yellow climbers are not common, so this is a rare find. As the name suggests, 'Garden Sun' climbs on a fence or pergola and seems to shine with its glowing yellow-apricot-peach blooms. Each individual flower is large, 3 to 4 inches across, with a double, cupped form. The petals themselves have kind of a wavy, scalloped appearance that adds lots of interest and texture. In cooler temperatures, I have seen some slight pink shading. All of this shining color is atop generous and healthy foliage.

Introduced in 2001 by Meilland (MEIvaleir).

Companions | Grow 'Garden Sun' up a fence, pillar, or arbor. Combine it with other pastels, whites, and roses with yellow flowers or stamens. Some good choices include 'Darlow's Enigma', 'Garden Delight', 'Morning Magic', 'Mutabilis', and 'Summer Sun'.

'Golden Fairy Tale'

Hybrid Tea (US); Shrub (Europe)

Bushy growth to about 3–4 feet

Yellow

I love yellow roses, and I am always thrilled to find good yellow roses that are disease resistant. 'Golden Fairy Tale' is one such rose, and it has been in my garden for a few years now. It just seems to keep getting better and better. I enjoy the complexity of the petals as the bloom opens—they have a very full, very double effect that could be called old-fashioned, a trait also rarely found in yellow roses. The blooms also have a blousy look to them, as they are not as tight as other quartered blooms. The rose has a light fragrance. In hottest conditions, the softness of the yellow color in the blooms can fade and I have seen some occasional black spot, but have always been impressed by how well it bounces back. Over the years 'Golden Fairy Tale' has scored high and consistently in rose trials—enough to include it in this book. It has been in international rose competitions, winning among other honors silver medals at Baden-Baden (2003), Kortrijk (2004), and the Federal Horticultural Munich (2005).

Introduced in 2005 by Kordes (KORquelda); also called 'Sterntaler'.

Companions | 'Golden Fairy Tale' is best in multiples of three or five in the border where the soft yellow blooms establish a presence rather than getting lost. Bring out the yellow color with other flowers that have a prominent stamen display or yellow coloration. Try 'Bajazzo', 'Escimo', 'Morning Magic', 'Mutabilis, and 'Oso Easy Cherry Pie'.

Disease resistance	55
Flowering	20
Fragrance	5
Total rating	80

'Golden Gate'

Disease resistance	60
Flowering	21
Fragrance	5
Total rating	86

Large-flowered climber

Climber to 8-10 feet

Yellow

You'll know by now that I'm partial to yellow roses in the garden and this climber is a beautiful yellow climber with full, large petals flowers that shine like the sun. The lemony fragrance is an added benefit, so place it in a spot in the garden where you can savor it. The growth habit is upright and tall but 'Golden Gate' could also be a free-flowing shrub swaying in the wind. I have never seen any black spot on this rose, which I think is quite an achievement for yellow roses.

Introduced in 2005 by Kordes (KORgolgat).

Companions | 'Golden Gate' is a true yellow that holds its color well. Because of this, I would choose varieties that play off that yellow with matching tones, such as 'Garden Delight' and 'Postillion'. It contrasts well with hot oranges and reds such as 'Brothers Grimm' and 'Summer Sun'. 'Escimo' has long yellow stamens that really pop when it is planted next to the yellow of 'Golden Gate'.

Disease resistance	48
Flowering	20
Fragrance	1
Total rating	**69**

Hybrid Tea

Upright growth to 5–6 feet

Bright, true red

One would think that the classic red rose would be just that—strong and symbolic, with a high-centered flower, wonderful fragrance, and long stems for cutting. Visitors often come into my garden looking for the classic red Hybrid Tea 'Mr. Lincoln', which they expect to have all these features. But much to their disappointment, I removed 'Mr. Lincoln' from the garden long ago due to its disease susceptibility.

I have included 'Grande Amore' in this directory despite the fact that it doesn't have the strongest disease rating because it is one of the best red roses available, and a good alternative to 'Mr. Lincoln'. I've been growing this rose in the high-disease-pressure regions of the East Coast and have seen some black spot, but growers on the West Coast sing its praises. 'Grande Amore' is one of the few Hybrid Tea roses to have been awarded the prestigious ADR Rose Award in 2005, and gold medals at the Den Haag Rose Trials (2004), Rose Hill Trials (2007), and at the Portland Rose Trials (2010),

so it is certainly performing well in many different regions.

Introduced in 2004 by Kordes (KORcoluma).

Companions | 'Grande Amore' grows very upright and is a good specimen for the middle to back of the border where you can cover up those bare legs if needed with companion plants. Complement 'Grande Amore' with other blends that pick up the red color in their buds like 'Caramella', 'Garden Delight', 'Poseidon', 'Postillion', and 'Tequila'.

'Heart Song'

Disease resistance	52
Flowering	26
Fragrance	3
Total rating	**81**

Hybrid Tea

Bushy growth to 3–4 feet

Red

I have been so pleased to discover another red rose with good disease resistance. 'Heart Song' has proven to be extremely healthy in such diverse regions as the Pacific Northwest and the South. The buds are very full and rounded—almost ball shaped, which open slowly opening to reveal unusual slightly frilled petals. The foliage is quite dark and glossy and the new growth is a reddish purple, which makes a nice background for the transparent red color of the blooms. For a red Hybrid Tea, this is a rose to make hearts sing.

Introduced in 2012 by Kordes (KORtrinka); also sold as 'Traumfrau'.

Companions | The bushy nature of this rose makes it suitable for the middle or front of the border. As a red, it can really stand on its own, but to complement the color, try some other varieties that share the same frilly textured petals, such as 'Alexandra Princesse de Luxembourg', 'Caramella', 'First Crush', 'Postillion', and 'Summer Romance'.

Disease resistance	60
Flowering	21
Fragrance	0
Total rating	81

Shrub

Bushy growth to about 4–5 feet

Red

As the name implies, this is a winning rose for any chemical-free rose garden. The single blooms have a true, constant red color and the plant is never out of bloom. The foliage is extremely healthy and deep green. One of the parents of this plant is the famed 'Knock Out', so you would expect all of the great characteristics of its heritage. 'Pink Home Run' (WEKphorn) is a color sport of 'Home Run' discovered in 2009 and introduced in 2010. Both varieties are fantastic as a rose hedge, providing great color all season long.

Introduced in 2006 by Weeks Roses; hybridized by Tom Carruth (WEKcisbako).

Companions | Whether you choose the red 'Home Run' or the 'Pink Home Run', you are going to have constant color and a bushy habit well suited to a rose hedge or border. If you want to complement with other roses, keep the forms simple and play off the red or pink with blends. For 'Home Run' try 'Brothers Grimm', 'Caramella', 'Poseidon', and 'Thrive! Lavender'. For 'Pink Home Run' try 'Easter Basket', 'Garden Delight', 'Marie-Luise Marjan', 'Savannah', and 'Stanwell Perpetual'.

'Innocencia Vigorosa'

Disease resistance	59
Flowering	20
Fragrance	3
Total rating	82

Floribunda

Compact, bushy, and spreading to about 3 feet

White

'Innocencia Vigorosa' is part of the Vigorosa series of roses bred by Kordes and Newflora specifically to be robust, disease-resistant, and floriferous. The very highlights of this breeding effort are combined in this assortment of roses. All of the Vigorosa roses have received ADR Rose Awards.

I was so struck by the health and vigor of this rose from the start that I selected it as one of the varieties for our Northeast Earth-Kind trials. The plants are vigorous in their growth and very healthy, with super shiny dark green leaves. I have never seen any black spot on this rose since I have had it in the garden. As beautiful as the foliage is, however, there is such a veritable sea of blooms produced on this plant that at many times during the growing season, you can't see the leaves at all. The individual blooms are the purest of white and about 2 inches across. They have about twenty petals, and open flat revealing a generous stamen presentation. The white of the petals and the gold of the stamens is a classic color match. It's like the most elegant white dress paired with gold earrings. The sheer number of blooms of 'Innocencia Vigorosa' boosts the fragrance.

The number of international awards this rose has garnered attests to its range. 'Innocencia Vigorosa' has received among other awards an ADR Rose Award in 2003, a gold medal at the Rome Rose Trials (2002), a Certificate at The Hague (2002), and a Certificate of Merit at the Australian National Rose Trials (2006).

Introduced in 2003 by Kordes (KORstarnow).

Companions | The growth habit of 'Innocencia Vigorosa' makes it suitable for the front of the border, planted along the pathway, or in a container. A mass planting would provide a stunning white rose hedge rarely without flowers. Because of its pure white color, virtually anything can color-match this rose. Here are some ideas to play off the bloom and the stamen color: 'Lemon Fizz', 'Oso Easy Cherry Pie', 'Mandarin Ice', 'Roemer's Hip Happy', and 'Ruby Ice'.

'Jane Bullock'

Shrub

Compact and bushy to 3–4 feet

Yellow

The wonderfully bright yellow color of the petals of this rose in contrast with the yellow-orange stamens makes me think of a bowl of oranges and lemons. This color combination will add cheer and brightness to any area of the garden. The flowers are prolific in their production and loosely formed, revealing the explosion of stamens and giving 'Jane Bullock' a more casual appearance in the garden.

Introduced in 2002 by Antique Rose Emporium.

Companions | This rose would be beautiful planted as a lovely yellow hedge, near the pathway where you can enjoy the fragrance, or in a pot. You can play off the yellow hues with orange roses such as 'Mutabilis' and 'Summer Sun'. Some other pinks would also make good companions, including 'Garden Delight', 'Roemer's Hip Happy', and 'Topolina'. Any red rose with yellow stamens would also pair well with 'Jane Bullock'.

Disease resistance	49
Flowering	22
Fragrance	4
Total rating	75

'Jasmina'

Disease resistance	54
Flowering	19
Fragrance	4
Total rating	77

Large-flowered climber

Climbing to 15 feet

Pink blend

'Jasmina' is one of the stronger climbers I grow in my garden. It is a fragrant variety that produces an abundance of blooms. I have heard the scent described as fruity, like apple, apricot, and pear. These pink flowers have a nodding effect that make them appear to be looking down in a friendly way. The overall effect of the color is a multitude of pinks combining to give a bright, pleasing appearance.

This roses won an ADR Rose Award in 2007, along with several gold and silver medals and fragrance awards in international trials, including the Australian National Rose Trials Certificate of Merit (2012) and a 2008 Kortrijk Rose Trials silver medal (2008).

Introduced in 2005 by Kordes-Newflora (KORcentex).

Companions | Plant this climber where it can complement the whites and pinks of companion roses like 'Beverly', 'Quietness', 'Souvenir de Baden-Baden', 'Summer Memories', and 'Thérèse Bugnet'.

Hybrid Kordesii (shrub)

Climbing to 10–12 feet

Pink

The Explorer Series collection from Agriculture Canada is a series of roses bred for hardiness (think of Canadian winters). 'John Davis' is a great example of the toughness of these hardy roses in the garden. If hardiness is not a concern in your garden, then grow this rose for its beautiful flowers and the fragrance. 'John Davis' can be grown on a fence or as a specimen in the back of the border. Reliable and consistent.

Introduced in 1977 by Felicitas Svejda and Agriculture Canada.

Companions | The color of this rose works well with a multitude of others. Consider other pinks, both lighter and darker, in 'Alexandra Princesse de Luxembourg', 'Eliza', 'Larissa', and 'Pomponella', or whites such as 'Summer Memories' to bring out the lighter tones.

Disease resistance	52
Flowering	20
Fragrance	0
Total rating	72

'Julia Child'

Disease resistance	50
Flowering	20
Fragrance	5
Total rating	75

Floribunda

Bushy and shorter, to about 3 feet

Yellow

One of the things I remember most about the great chef and personality Julia Child was her love of butter. This butter-yellow rose is thus a most appropriate rose to carry her name, and in fact the rose was chosen by Julia Child herself. The stunning yellow blooms are produced in abundant clusters. They are cupped, and have a ruffled, old-fashioned appearance. The fragrance is described as strong and licorice. I have seen some occasional black spot in extreme conditions, but this rose seems to hold strong no matter what—much like its namesake. This rose was declared an AARS winner in 2006.

Introduced in 2006 by Weeks Roses; hybridized by Tom Carruth (WEKvossutono).

Companions | The bushy habit of 'Julia Child' means that it sits well in the middle or front of a mixed border. Plant it where you can enjoy the fragrance; it's even suitable for a container. The yellow of 'Julia Child' can be complemented by other roses that have prominent stamen displays, yellow blends, and whites. Try it with 'Bajazzo', 'Easter Basket', 'Garden Delight', 'Mutabilis', and 'Roemer's Hip Happy'.

Disease resistance	**58**
Flowering	**29**
Fragrance	**0**
Total rating	**87**

Shrub

Short and bushy to about 3 feet

Pure red

'Kardinal Kolorscape' is one of the red roses in the Kolorscape series and, to my mind, one of the prettiest. It has a semi-double flower form with about ten to fifteen petals that generally stay in more of a cup shape. The blooms and their color seem almost electric, and they really glow in the garden. 'Cherri Kolorscape' and 'Milano Kolorscape' are other reds in the series.

Introduced by 2012 by Kordes (KORsixkono).

Companions | 'Kardinal Kolorscape', like all the roses in the series, works well planted in the front of the border, along a pathway, or in a mass planting or rose hedge. In the rose garden it partners well with any other hot colors, deep pinks, or whites. Try it with 'Caramella', 'Dee-Lish', 'Mutabilis', 'Oso Easy Cherry Pie', and 'Raspberry Kiss'.

'Karl Ploberger'

Shrub

Upright and tall to 4–5 feet

Yellow

This is a widely adapted yellow rose that deserves to be grown in more gardens. Its namesake is a gardening guru in Australia, but the most captivating part is not its name—it is indeed the bloom. When I can take my nose out of the bloom for a few seconds (it is so fragrant), I enjoy the subtle color display of yellow with lighter outer petals. It almost is as if the bloom opens with the lighter petals and then reveals the darker yellow inside, and in a reverse of the usual it gets darker in color with age. The petals themselves have a wavy, scalloped effect that adds depth and texture. I have seen some occasional black spot on this rose where there is high disease pressure, but it is pretty resilient and 'Karl Ploberger' has scored well in trials over the years. With apologies to Karl, perhaps if this rose had a different name, it would gain popularity outside Australia.

Introduced in 2008 by Kordes.

Companions | 'Karl Ploberger' grows upright and can become quite tall, so put it in the middle to back of the border. If you have a place where you can enjoy the fragrance, all the better. The colors in this rose are a creamy white to yellow, and I love to play off yellow colors with other plants that have brilliant stamen displays. Any yellows, yellow blends, and whites will complement this rose, as will hotter colors and pastels. Try it with 'Darlow's Enigma', 'Garden Delight', 'Innocencia Vigorosa', 'Macy's Pride', and 'Mutabilis'.

Disease resistance	50
Flowering	19
Fragrance	7
Total rating	76

Shrub

Bushy and taller to about 4–5 feet

White

'Kew Gardens' has a simple, single flower form that is a far cry from the multi-petaled form of many Austin roses. However, these simple flowers are produced in such large, almost hydrangea-like clusters that the whole effect is greater than that of any multi-petaled variety. The bud color is a soft peachy apricot, yet the flowers open wide to a brilliantly white single bloom with a stamen display that looks like golden fireworks casting a yellow glow back down onto the white background. 'Kew Gardens' is one of the more disease-resistant varieties, and I love the flexibility it offers in the landscape. It was the winner of a Gold Standard at the 2010 Gold Standard Rose Trials.

Disease resistance	52
Flowering	24
Fragrance	1
Total rating	77

Introduced in 2012 by David Austin Roses (AUSfence).

Companions | 'Kew Gardens' makes a beautiful specimen plant in the border, in a container, or planted in mass as a large hedge. The possibility of color combinations here is nearly endless. I like to play off the subtle yellow-apricot tones seen early in the bloom and also use some white blends. Try it with other David Austin roses in these colors, or 'David Rockefeller's Golden Sparrow', 'Karl Ploberger', 'Postillion', 'Winter Sun', and 'Yellow Submarine'.

David Austin English Roses

The David Austin English Roses have enjoyed great success for some time now and many gardeners find them enchanting, with good reason. If this book were solely about roses with abundant bloom, luxurious flower form, and fragrance, its pages would be filled with many of these varieties. Their blooms are famously enjoyed by anyone lucky enough to see one and to breathe in its scent. These roses typically combine the best qualities and appearance of old garden roses, but with the bonus of repeat bloom. In my region, where humidity levels are high during the growing season, these roses can get disease. In other areas, however, such as the West Coast, gardeners enjoy these roses will little to no chemical intervention. If you live in an area with high disease pressure, you will have to decide if having these stunning roses is worth the risk of the foliage developing fungal disease. I certainly have enjoyed many of the latest introductions from David Austin, particularly because they don't look, smell, or bloom like any other roses.

In addition to 'Kew Gardens', other varieties in the English Rose collection that I have found display greater disease resistance are 'Benjamin Britten', 'Crocus Rose', 'Heritage', 'Huntington Rose', 'Queen of Sweden', 'Skylark', 'Sophy's Rose', 'Strawberry Hill', 'The Generous Gardener', 'The Lady's Blush', and 'The Shepherdess'.

'Knock Out'

Disease resistance	57
Flowering	25
Fragrance	0
Total rating	82

Shrub

Bushy growth to 3–4 feet

Cherry red

The original 'Knock Out' is one of the rose legends of our time. This single variety has convinced people to start growing roses again, and to grow them with success and without chemicals. 'Knock Out' sometimes receives criticism for its flower form or lack of fragrance. However, those qualities were not what the hybridizer was seeking—he was specifically aiming for disease resistance, and for that effort I thank him. In addition to the beautiful, disease-free foliage, 'Knock Out' never seems to be out of flower. The ability to produce such multitudes of these cherry-red blooms gives any landscape gorgeous non-stop color. 'Knock Out' has been given the Earth-Kind designation, as well as being an All-American Rose Selection in 2000 and an ADR Rose Award in 2002.

The Knock Out family of roses includes the original 'Knock Out' along with 'Blush Knock Out', 'Double Knock Out', 'Pink Double Knock Out' 'Pink Knock Out', 'Rainbow Knock Out', and 'Sunny Knock Out'.

Due to the great marketing efforts behind these roses they are well known to home gardeners. Each one of these roses deserves a page in this book. My personal favorite is 'Blush Knock Out', with 'Pink Knock Out' a close second; both are color sports of 'Knock Out'.

Introduced in 2000 by Conard-Pyle; hybridized by Will Radler (RADrazz).

Companions | Whether planted as a single plant, in multiples, or in a mass grouping, with this rose you will achieve a constant color presence in your garden. I have enjoyed 'Knock Out' most when used in large plantings. I do find the color of 'Knock Out' harder to blend with a lot of colors, and one of the things I love to do in the rose garden is experiment with the synergy of colors to see how the color of one rose can influence that of another. It can be paired with other color blends or with roses that have bud color in the same red color family. Some good suggestions are 'Caramella', 'Garden Delight', 'Poseidon', 'Raspberry Kiss', and 'Thrive! Lavender'.

Disease resistance 55
Flowering 25
Fragrance 3
Total rating 83

Large-flowered climber

Climbing to 10–12 feet

Yellow

People who have visited my garden always stop to ask what this rose is and they are amazed to hear that I have only been growing it for a few years. 'Kordes Moonlight' has grown easily and vigorously for me from the get-go, covering a tall fence within the third year. The foliage is the glossiest of greens, making it a beautiful plant even if it didn't bloom. But boy does it bloom! I think "moonlight" is a little misleading as I was expecting this rose to bloom white. But I was delightfully surprised at the multitude of colors this rose produces. The buds have a darker orange and pink to them, then the flower opens to a strong yellow. As the blooms age they fade to a lighter yellow. All in all, the rose provides a wonderful painting of colors.

Introduced in 2004 (Germany) and 2008 (United States) by Kordes (KORklemol).

Companions | You can have a lot of fun with the color matches for 'Kordes Moonlight', especially with the deep colors of the orange and deeper pink-red buds. Try it with 'Garden Delight', 'Mandarin Ice' and 'Summer Sun'. For pinks, try 'Pomponella' and 'The Fairy'. For whites, try 'Polar Express' or 'Summer Memories'.

'KOSMOS'

Disease resistance	**60**
Flowering	**25**
Fragrance	**5**
Total rating	**90**

Floribunda

Shorter and bushy to 3 feet

White to cream

'KOSMOS' has become a cornerstone of consistency in my own rose gardens both in New York and in Maine. Another in the line of Fairy Tale roses from Kordes, this is a modern, robust, and resilient rose that gives a nod to an old-fashioned flower form. 'KOSMOS' has the highest rating for disease resistance, always presenting a covering of lush, green foliage from the top of the bush to the very bottom.

The flowers have a classic bud form and the buds often show a hint of pink or peach tints or even darker pink hues on the edges as it cracks open the sepals. The flower opens to reveal a many-petaled bloom, but the petals are loosely packed, which gives the bloom substance and fullness. If 'KOSMOS' has a flaw, it's that the blooms are sometimes too large for the stems to hold upright. For the average rose gardener I would call this a happy problem. The open blooms can have a little bit of a ruffled edge and even more have a creamy center. All in all, this is a top-rated rose with the greatest of disease resistance. It was awarded an ADR Rose Award in 2007 and First Prize at the Hradec Rose Trials (2009).

Introduced in 2006 by Kordes (KORpriggos).

Companions | With its smaller, bushy habit, 'KOSMOS' is great planted in masses as a rose hedge, near the front of the border, or along the pathway or drive. It also looks good in a pot where you can observe the gently nodding flowers. The white color allows for maximum color choices—it could partner with almost any rose. To narrow it down a little, I suggest you plant it with other creams, apricots, and peaches like 'Eifelzauber' or 'Mandarin Ice' with its white reverse. A good partner in red would be 'Black Forest Rose'. Pinks and whites are a classic combination to try, such as 'Beverly' and 'Wedding Bells'. 'Ruby Ice' has a white reverse that would also be a great pairing.

Disease resistance	60
Flowering	19
Fragrance	0
Total rating	79

Shrub

Bushy growth to 3–4 feet

Coral-pink

Another great shrub rose with a color that is unlike any other, 'Lady Elsie May' has extremely healthy, glossy foliage. The strength of this rose is its consistency; once it starts flowering, it doesn't stop until frost. Whoever Lady Elsie May could have been, perhaps she kept going without a care in the world, just like her namesake rose. The blooms have about ten to fifteen petals, and open flat revealing a subtle patch of stamens in the middle.

Introduced in 2002 by Noack (NOAelsie).

Companions | 'Lady Elsie May' gives great color all the time. For this reason, I've found it to be really effective planted in numbers, typically three, five, or more. Planted in mass or as a hedge it can also be really stunning. Try it in a container as well. The wonderful color can be a bit tricky to match, but try and complement it with other apricots, peaches, lighter pinks, reds, and whites. I suggest 'Larissa', 'Mother of Pearl', 'Mutabilis', 'Savannah', and 'Thrive! Lavender'.

'Lady Pamela Carol'

Disease resistance	53
Flowering	21
Fragrance	2
Total rating	76

Shrub

Strong, upright growth to 5–6 feet

Yellow

This rose has become a favorite yellow in my garden. It is a strong, vigorous grower and definitely will add presence to a mixed border. The foliage is a nice dark green, and I have rarely seen any signs of disease on it. The flowers are a wonderfully soft yellow, one of those shades that is very calming and pleasing. The plant and the color of the blooms can hold their own in the border or blend in nicely with a soft pastel color palette. The flowers are typically 3–4 inches across and open flat and loosely arranged. There is something about this variety that seems to be generous in its spirit.

Introduced in 2006 by the Antique Rose Emporium.

Companions | The height of 'Lady Pamela Carol' means it can act as a specimen plant in the mixed border, or blend nicely near the back of the border with other pastel colors. Try it with 'Darlow's Enigma', 'Morning Magic', 'Mutabilis', 'Old Baylor' and 'Thérèse Bugnet'.

Disease resistance	60
Flowering	20
Fragrance	3
Total rating	83

Large-flowered climber

Climbing to about 15 feet

Deep pink

One of the best of the best, and widely adaptable, 'Laguna' is a great addition to the garden on so many levels. First of all, it is a gorgeous plant. 'Laguna' has leaves of the deepest green that are always glossy and healthy. The plant is covered head-to-toe with this dense, prolific foliage and I have never seen any black spot on it in all the years that I have been growing it. The growth habit is upright and strong—even if this plant didn't have flowers it would be a great addition to a fence or pergola. But, as luck would have it, 'Laguna' does bloom, with flowers that are a deep pink or even wine color. They are very fully ruffled in an old-fashioned way and are produced in clusters of five to seven flowers—an explosion of flowers that nearly covers the entire plant.

Although I don't often rave about fragrance, this is one to talk about. It has been described as intoxicating and fruity, like a spicy perfume. In my seaside garden in Maine, this is the rose I choose to cover a long white fence. The lush green of the foliage against the white fence is classic. I look forward to the wine-colored blooms releasing their perfume with the sea breezes. What about Maine winters you ask? Yes, this rose has proven hardy. Whether you have a farm, a small courtyard, or a lamp post, this is a rose to consider for every garden. It was awarded the ADR Rose Award in 2007, and has won awards for its appearance and fragrance, among them silver medals in Baden-Baden, Geneva, and The Hague.

Introduced in 2008 by Kordes (KORadigel).

Companions | Grow 'Laguna' over a strong structure or fence, where its deep color makes a good foil for any roses in front of it. Try 'Brilliant Veranda', 'Caramella', 'Eifelzauber', 'Poseidon', and 'Mother of Pearl'. Some of the bigger bushes like the Noisettes would also be stunning set in front of 'Laguna'. Other climbers to grow with it include 'Awakening', 'Jasmina', 'Nahema', 'New Dawn', and 'Peggy Martin Survivor'.

'La Marne'

Disease resistance	51
Flowering	18
Fragrance	1
Total rating	70

Polyantha

Upright and erect to about 3–5 feet

Pink and white blend

'La Marne' is one of the most consistent roses in my garden, and it has been given the Earth-Kind designation. That's why I am often confused when visitors walk right by it. 'La Marne' is always in color and shows little to no signs of disease. Perhaps it's because 'La Marne' has a simple charm. I suggest planting this variety in numbers to make people stop and take notice. The blooms are small—about 1 inch in diameter, pink with a creamy white blend. The petals themselves have a somewhat pointed and scalloped edge that gives the overall effect more texture and interest. Charming. Although this rose does exceptionally well in hotter climates, I've had success in zone 6 where it grows a little less tall but otherwise performs very well.

Introduced in 1915 by Barbier Frères & Compagnie.

Companions | 'La Marne' does well in a container or planted in groups or in a hedge. The pink and white blend gives you the opportunity to try a lot of combinations of other pinks and whites with different flower forms, shapes, and sizes. I suggest 'Carefree Beauty', 'Eliza, 'Lion's Rose', 'Polar Express', and 'Wedding Bells'.

Hybrid Tea
Stout and upright to 4 feet
Creamy white

For a Hybrid Tea, 'La Perla' has one of the highest black spot resistance ratings and has been given an ADR Rose Award (2009), among other prizes. It grows well in areas with warmer summers, but I have never seen any type of leaf spot on this rose since I have been growing it in the Northeast. The leaves are dark green, glossy, and very thick and substantive, making a perfect setting for the creamy white flowers. 'La Perla' is a completely fitting name for this rose—it is like a perfect shiny pearl set on a bed of green.

The blooms are usually borne singly (as in most Hybrid Teas) and unfurl from a high-centered bud. The open bloom petals also seem to have substance and a firmness to them. They open in a wide, blowsy way to become 4–5 inches wide. Depending on the temperature, there may also be the softest yellowing in the center of the bloom. This coloring along with the thick petals gives the bloom a lot of texture and depth. Stunning.

Introduced in 2008 by Kordes (KORpenparo).

Companions | 'La Perla' has a very upright and stout, typical Hybrid Tea growth habit. It is well suited to the middle or closer to the front of the border where it can shine surrounded by complementary and contrasting colors. Try it with 'Easter Basket', 'David Rockefeller's Golden Sparrow', 'Oso Easy Cherry Pie', 'Ruby Ice', and 'Wedding Bells'.

Disease resistance	60
Flowering	18
Fragrance	2
Total rating	80

'Larissa'

Disease resistance	**60**
Flowering	**26**
Fragrance	**0**
Total rating	**86**

Floribunda

Bushy and spreading to 3–4 feet

Pink

Much as I promise myself not to have favorites, this rose is a favorite. When I first planted 'Larissa', it was barely a rooted cutting. It went into the ground so small that I thought there was no way it could survive. Not only did it survive, it thrived, sending up strong shoots and quickly developing a presence in the garden. In all my years of growing this rose, I have not seen any black spot on it. It has been awarded a number of medals in international rose competitions, including a gold medal in Kortrijk (2007, 1010) and an ADR Rose Award in 2008.

For such a tough rose, it has super-charming little blooms. They are old fashioned and quartered in their appearance. Combine twenty, thirty, or fifty of these little blooms together and you get quite a panicle of bloom that makes you forget the fact that they aren't fragrant. And to top things off, the petals fall very cleanly from the spent blooms, leaving the plant tidy at all stages. A colleague who witnessed 'Larissa' from the very first planting often says, "When in doubt, plant 'Larissa'." I couldn't agree more.

Introduced in 2007 by Kordes (KORbaspro).

Companions | 'Larissa' in a mass planting is show stopping. The short, bushy nature of the plant makes it suitable for a container. In the border, place it toward the front where you can enjoy the charm of the blooms. The light pink color of the blooms fades as it ages, so any roses in pastels or white are complementary. Try 'Alexandra Princesse de Luxembourg', 'KOSMOS', 'Lion's Rose', 'Pink Enchantment', and 'Summer Memories'.

Disease resistance	59
Flowering	27
Fragrance	0
Total rating	86

Shrub

Bushy, rounded growth to 3 feet

Yellow

'Lemon Fizz' is the yellow rose in the Kolorscape series. It has a simple flower form, with five petals that open flat and reveal a lovely cluster of stamens. I have seen the stamens present with colors that range from a wonderful orange color to golden-yellow. The yellow color of the blooms is very fade-resistant.

Introduced in 2012 by Kordes (KORfizziem).

Companions | 'Lemon Fizz', like all the roses in the Kolorscape series, works well planted in the front of the border, along a pathway, in a mass planting, or as rose hedge. If you want to use it in the garden with mixed plantings, 'Lemon Fizz' has such a bright, non-fading yellow that it stands up to any other hot colors. I suggest pairing it with orange 'Brothers Grimm' and 'Garden Delight', red 'Thrive!', or white 'Darlow's Enigma' and 'Innocencia Vigorosa'.

'Lena'

Disease resistance — 55
Flowering — 22
Fragrance — 0
Total rating — 77

Shrub

Bushy and spreading to about 3–5 feet

Pink blend

'Lena' and 'Ole' were both introduced as part of the Northern Accents series of roses by Kathy Zuzek, bred specifically for cold hardiness. I have enjoyed these plants so much in the garden, watching as they have grown to large shrubs with prolific displays of charming flowers. Both of these varieties scored so consistently high in rose trials that I placed them in the Northeast Earth-Kind observational trial, where they both did very well. They both deserve to be grown more widely, especially in cold-winter areas.

Hybridized in 2007 by Kathy Zuzek; introduced by Bailey Nurseries (BAIlena).

Companions | 'Lena' can be planted as specimen plant in the border, in multiples, or as a rose hedge. You can play off the pink bud color—white and pink blooms that have the same nostalgic feeling would complement it well. I recommend planting 'Lena' with 'Belinda's Dream', 'Blush Noisette', 'Carefree Beauty', 'Pink Drift', and 'Thérèse Bugnet'.

Floribunda

Bushy and upright to 4–5 feet

White

When describing this rose, I can't get the word joy out of my mind. The enjoyment of this rose is in discovering all of its pleasing characteristics. One of the beautiful color traits of 'Lion's Rose' is a tendency for the blooms to take on cream tones with a slightly pinkish or apricot center. The blooms have a lovely, old-fashioned look, opening flat to reveal multiple petals. They are produced in clusters of seven to nine blooms on a single stem. 'Lion's Rose' is always eager to flower. The foliage is lush green and glossy. Every rose gardener should have the chance to love this rose, which is indeed as strong as a lion. Cherish the joy—this rose brings pleasure and delight in its ease, health, vigor, and beauty.

'Lion's Rose' is one of the Fairy Tale series, and it has been awarded an ADR Rose Award (2002), a gold medal in the Gold Standard Rose Trials (2006), and many other medals in international competitions.

Introduced in 2007 by Kordes (KORvanaber); also sold as 'Lion's Rose Fairy Tale'.

Companions | With its taller habit, 'Lion's Rose' fits in the middle to back of the border. Mix it among other creams, apricot, peach, and red. Try pale 'Caramella' and 'Eifelzauber', or red 'Fiji'. Pinks and whites create classic combination as well, like 'Alexandra Princesse de Luxembourg' and 'Wedding Bells'.

Disease resistance	58
Flowering	26
Fragrance	4
Total rating	88

'Lupo'

Disease resistance	**52**
Flowering	**21**
Fragrance	**0**
Total rating	**73**

Miniature

Dense and bushy to 3 feet

Purple to carmine-red

Don't let the description "miniature" dictate your idea of this plant. 'Lupo' is a very dense, compact bush with a profusion of charming little flowers borne in clusters that give the effect of a mass of color. Each individual flower has about five to eight petals that open flat to reveal a pretty little pouf of yellow stamens set off by a lighter (sometimes white) eye. The flower clusters are extremely weather-resistant and produce abundant hips in the fall. The foliage is deep green, dense, glossy, and extremely healthy.

This rose has a lot of vigor and has won many international awards, including an ADR Rose Award (2007), with strong performances in international rose trials including The Hague Rose Trails (Certificate of Merit, 2006), Geneva Rose Trials (Certificate of Merit, 2008), Belfast Rose Trials (Best Miniature or Patio Rose, 2008), and the Australian National Rose Trials.

Introduced in 2007 by Kordes (KORdwarul).

Companions | Plant 'Lupo' with some contrasting colors and golden-yellows such as 'Carefree Sunshine', 'Centennial', 'David Rockefeller's Golden Sparrow', 'Lady Pamela Carol', and 'Solero'.

Disease resistance	57
Flowering	25
Fragrance	4
Total rating	86

Shrub

Bushy, tall growth to 5 feet

White

Selected for the centennial celebration of the department store Macy's, this rose is such a strong performer I suspect it will be still be around for the bicentennial celebration. 'Macy's Pride' grows upright and strong. The buds have a lemony yellow quality with a kind of luminescence, like an inner glow. The petals unfurl to a somewhat lax habit of a semi-double bloom that evolves to a steady white.

These blooms are sometimes produced singly but often in wonderfully profuse clusters of three or five blooms that are so abundant they can proudly cover the glossy, green foliage in a sea of white.

'Macy's Pride' is part of the Easy Elegance collection and its parentage includes the beauty of the English rose 'Graham Thomas' and the strength of 'Carefree Beauty'. With such an illustrious heritage, it's no wonder that this is such a strong, easy, and satisfying rose to grow.

Introduced in 2003 by Bailey Nurseries; hybridized by Pink Lim (BAlcream).

Companions | The growth habit of 'Macy's Pride' is tall and upright so it makes a good background to any complementary colors. Try some companions of yellow to highlight the lemony bud color, such as 'Alister Stella Gray', 'David Rockefeller's Golden Sparrow', 'Golden Fairy Tale', 'Karl Ploberger', and 'Postillion'.

'Mandarin Ice'

Disease resistance	**60**
Flowering	**30**
Fragrance	**1**
Total rating	**91**

Floribunda

Bushy growth to 4 feet

Mandarin-red with a silvery white reverse

The original German name of this rose, 'Planten un Blomen', is spot on because if you plant it, it will bloom. In fact, this is a flowering powerhouse of a rose.

The color is vibrant and exciting and I have seen extremely large clusters (up to fifty blooms) on each shoot. With the lighter reverse, the color effect is a real conversation piece. The foliage is incredibly healthy, deep green, and glossy.

Introduced in 2008 by Kordes as 'Planten un Blomen' and in the US in 2011 as 'Mandarin Ice' (KORplunblo).

Companions | The vivid color of 'Mandarin Ice' mixes well with other hot colors in the garden. Try it with other oranges, reds, and yellows such as 'Garden Delight' and 'Summer Sun'. The pale reverse on the blooms suggests some companion whites as well; I recommend 'Darlow's Enigma', 'KOSMOS', and 'Polar Express'.

Polyantha

Shorter and bushy to 3 feet

Pink

'Marie Daly' is a modern color sport of the rose 'Marie Pavie', a Polyantha that has been around since 1888. Both are worth mentioning for their obvious charms. As Polyanthas (poly meaning many), they have many clusters of small blooms that cover that plant with color and fragrance. Disease resistance and toughness are a given—'Marie' has stood the test of time. The lack of thorns, or few thorns on this rose, also make this a good variety in areas with heavy foot traffic. 'Marie Daly' has been designated as an Earth-Kind rose.

Companions | 'Marie Daly' and 'Marie Pavie' are both suitable along pathways, in containers, or placed in a mass planting. The pink color of 'Marie Daly' is flexible and can work with many other colors. With the softness of the flower, I enjoy combining it with other roses that share the same quality, such as 'Carefree Beauty', 'Cubana', 'Ducher', 'Marie-Luise Marjan', and 'Peach Drift'.

Disease resistance	54
Flowering	25
Fragrance	3
Total rating	82

'Marie-Luise Marjan'

Disease resistance	**54**
Flowering	**25**
Fragrance	**9**
Total rating	**88**

Hybrid Tea

Bushy, upright growth to 4–5 feet

Creamy white

'Marie-Luise Marjan' has been available for some years and so is not part of the new movement toward breeding disease resistance, but it is included in this book because of its garden-worthy qualities. It was awarded a gold medal in Geneva in 1997 and a silver medal in Magdeburg in 1999, and since then it has stood the test of time. Most Hybrid Teas are upright and sometimes stiff, but I find that this variety wants to grow more bushy and shrublike. I don't really mind how it grows, just as long as it keeps on consistently producing those beautiful, fragrant flowers. The buds are high-centered and sometimes have a tint of pink or apricot to them. As they unfurl they reveal a loosely formed flower composed of petals with a waviness about them. Fully opened blooms offer a surprise of long stamens, like long eyelashes begging for you to take notice.

The real complexity of this rose I think is in the subtle shades of color. Depending on the temperature, it can display hints of pink, yellow, or peachy apricot on the creamy white petals. The effect is subtle and lovely. The fact that 'Marie-Luise Marjan' is also fragrant helps to boost its scores during evaluations, and the occasional touches of black spot are therefore forgivable.

Introduced in 1999 by Kordes (KORfinger).

Companions | Because of this variety's somewhat shrubby growing habit, I think it is best placed in the middle or toward the front of the border, or raised up in a container so the fragrance can be appreciated. The soft colors of its blooms can be complemented by other plants with soft pastel and white flowers. I suggest pairing it with 'Alexandra Princesse de Luxembourg', 'Eliza', 'Peach Drift', 'Savannah', and 'Winter Sun'.

Floribunda

Upright and bushy to 5 feet

Apricot-salmon

I have been growing this rose for some time now and it continues to shine year after year. The flowers are wonderfully subtle, with each individual bloom measuring about 2 inches across and quite flat with frilly edges. I like to think of the color as orange sherbet that fades to lighter salmon-pink. The best thing is that 'Michel Bras' produces huge panicles of these flowers set well above the foliage, ready to give you a smile as you pass by. The foliage is glossy and nicely mid-green.

Introduced in 2002 by Delbard (DELtil).

Companions | The unusual color of 'Michel Bras' can help you to create some unique color combinations in the rose garden. I love orange and pink together so like to plant 'Eliza' behind it so that the taller pink rose reaches up and over the sherbet orange. 'Quietness' is another nice pink to mix with the color. 'Cream Veranda' echoes the apricot tones in a lighter shade. Yellows would also blend beautifully and highlight the stunning stamens of this rose—I suggest trying it with 'Sunny Sky'—and white 'KOSMOS' would be lovely as well.

Disease resistance	53
Flowering	25
Fragrance	4
Total rating	82

'Miracle on the Hudson'

Disease resistance	**55**
Flowering	**22**
Fragrance	**1**
Total rating	**78**

Shrub

Bushy growth to 3–4 feet

Red

A great rose can be named to represent a heroic event. I was first introduced to this rose by my dear friend Pat Henry of Roses Unlimited Nursery in South Carolina. Pat chose the name after the famed landing of a plane in 2009 on the Hudson River by Captain Chesley Sullenberger, described as one of the most miraculous and skillful emergency landings in aviation history. Such an event calls for a suitable rose to commemorate the story and this is a great choice. I have enjoyed it in my garden ever since. Because 'Miracle on the Hudson' is bushy and shrublike in its habit it makes a great rose hedge. The blooms are a dark red color with even deeper, velvety tones that add luminosity. It displays the same qualities as its parent 'Home Run' and is healthy and carefree in every way. 'Miracle on the Hudson' will be as steady in your garden as Captain Sullenberger was at the controls.

Introduced in 2006 by Robert Rippetoe; also sold as 'Bartholomew'.

Companions | 'Miracle on the Hudson' is excellent for mass plantings but it is also suitable for the middle of the border or even a generous container. The deep colors of this bloom are a wonderful complement with other saturated colors, red blends, apricot-peach, and whites. I recommend planting it with the rose partners 'Caramella', 'Mutabilis', 'Poseidon', and 'Thrive! Lavender'.

'Morning Magic'

Disease resistance	**58**
Flowering	**20**
Fragrance	**2**
Total rating	**80**

Large-flowered climber

Tall and upright to about 7–8 feet

Light pink

I have grown this rose both on a fence and in the rose garden near the back of the border and enjoyed both displays. Hybridized by the breeder of 'Knock Out', 'Morning Magic' has the great disease resistance of its famous hybridizer. The healthy foliage is a superb backdrop for the porcelain blooms. The flowers open flat to reveal their delicate pink color, shaded with white and displaying yellow stamens in the center. The vigor of this plant and the color of the flower complement any rose garden.

Introduced in 2008 by Conard-Pyle; hybridized by Will Radler (RADmor).

Companions | 'Morning Magic' can be grown as a wonderful specimen plant near the back of the border, massed as a rose hedge, or it can be trained as a mannerly climber on a fence or post. The soft pink and creamy white color of the flowers blends well with other pinks and whites. You can also echo the yellow stamen display by matching with other yellows. Try it with 'David Rockefeller's Golden Sparrow', 'Larissa', 'Macy's Pride', 'Polar Express', and 'Yellow Brick Road'.

'Mother of Pearl'

Disease resistance 53
Flowering 26
Fragrance 2
Total rating 81

Grandiflora

Tall and upright to 6 feet

Apricot and peachy pink

The color of this rose is a real favorite with garden visitors. When 'Mother of Pearl' is in full bloom, people flock to take pictures of it or to have their pictures taken with it. The flowers have a classic form and are produced in great clusters—I have often seen eight to twelve blooms per stem. Some people smell a fragrance but the real attraction of 'Mother of Pearl' is its tall presence in the garden along with an abundance of blooms. 'Mother of Pearl' has consistently scored well in trials, and I have also found this rose to be very resistant to disease in my own garden.

Introduced in 2006 by Conard-Pyle; hybridized by Meilland (MElludere).

Companions | Have fun mixing 'Mother of Pearl' with lavender roses like 'Poseidon', orange-peaches such as 'Caramella', pinks such as 'Alexandra Princesse de Luxembourg' and 'Pink Enchantment', and whites like 'Lion's Rose'.

Disease resistance	53
Flowering	30
Fragrance	0
Total rating	83

China

Bushy upright growth to 6–7 feet

Yellow and orange to dark crimson

I hesitated putting this rose in the directory because it doesn't represent the modern roses that have been bred toward disease resistance. But 'Mutabilis' is one of those heritage varieties that has survived over the centuries without any chemical treatments. I grow it in my New York garden even though it's a China rose that really thrives in the heat.

'Mutabilis' is an Earth-Kind variety, and the program describes the rose well: "'Mutabilis' was introduced prior to 1894 and is one of the most famous and beloved of the old garden roses. Amazing medium-sized single blossoms that pass through three distinct color phases (hence the name 'Mutabilis', since the blooms "mutate" in color) beginning with yellow, changing to pink, and finally to crimson." These constantly changing colors are really a wonder. You can walk out into your garden and notice a particular bloom on this plant in one color. Walk out later on and you might notice the same bloom has deepened in color or changed entirely.

'Mutabilis' is also called the butterfly rose because its blossoms look like brightly colored butterflies that have landed on the bush. This large, attractive shrub was named 2005 Earth-Kind Rose of the Year: it is easy to grow and has great heat tolerance, making it well suited for growing in the South. Be sure to give it plenty of room to grow.

Companions | This rose can grow to be quite large. In the mixed border, it can hold its own near the back. In a mass planting or hedge, it can be a stunning, ever-changing live painting in your garden. Play off the colors to accentuate them even more. Try it with 'All the Rage', 'Coral Drift', 'David Rockefeller's Golden Sparrow', 'Flamenco Rosita', 'Lady Pamela Carol', and 'Mother of Pearl'.

'My Girl'

Disease resistance	**55**
Flowering	**26**
Fragrance	**0**
Total rating	**81**

Shrub

Medium, upright growth to about 4 feet

Deep pink

'My Girl' is part of the Easy Elegance collection. These roses claim "to deliver everything we love about roses—the color and beauty—without the fussy maintenance roses used to require." My experience with the Easy Elegance roses is that this is true. 'My Girl' has a very healthy, deep-green, glossy foliage that is a perfect background to the vivid pink blooms. They are almost a glowing lipstick pink. The flowers have about twenty to thirty petals and open up in a somewhat open fashion, holding the same solid color from start to finish. The petals themselves have a little ruffle to give texture and depth.

Introduced in 2008 by Bailey Nurseries (BAIgirl).

Companions | With such a striking color 'My Girl' can certainly get noticed in the middle of a mixed border. The vivid color holds a lot of possibilities for partnering with other roses, including some lavenders, oranges, pastels, and blends. I suggest planting it with 'Garden Delight', 'Easter Basket', 'Mutabilis', 'Poseidon', and 'Thrive! Lavender'.

Disease resistance	52
Flowering	21
Fragrance	7
Total rating	80

Large-flowered climber

Climbing to 9–12 feet

Light pink

I didn't know too much about the Delbard roses from France until I was doing research to diversify the collection of roses at the Peggy Rockefeller Rose Garden in 2007. Pat Henry from Roses Unlimited told me that this was one to try and she was certainly right. 'Nahema' is a sort of sleeper variety that not many people know about. It grows quite abundantly and produces beautiful, globular clusters of the softest pink. The blooms are very large—up to 4 inches across—and their cup shape holds such a delicious fragrance that you just want to drink it.

Introduced in 2006 by Delbard (DELeri).

Companions | Plant 'Nahema' on a structure where you can enjoy the fragrance. The color is easy to combine with other plants in the garden. I like to play off pink tones with white roses that are fragrant. Try 'Beverly', 'Blush Noisette', 'Darlow's Enigma', 'First Crush', 'Marie-Luise Marjan', 'Summer Memories', and 'Thérèse Bugnet'.

'Nastarana'

Disease resistance	51
Flowering	22
Fragrance	8
Total rating	81

Noisette

Tall, bushy growth to 5–6 feet

White

For a change I will start by describing the fragrance of this rose. 'Nastarana' has always won people over with its fragrance; it is what I would describe as licorice or anise. Looking at the introduction date of this plant should tell anyone that it is a pretty strong plant that has been around a long time. As with most Noisette roses, it blooms in large clusters of small flowers. In bud form, the color is a precious pink, opening to white. The combination of buds and mature flowers on the plant is stunning.

Discovered in 1879 by Pissard; also sold as 'Persian Musk Rose' and 'Pissardii'.

Companions | This is a great specimen plant on its own, or it can be planted in the back of the border where it has some room. I have planted it with other pinks to play off that bud color. Try 'Alexandra Princesse de Luxembourg', 'Eliza', 'Fortuna Vigorosa', 'Larissa' and 'Pomponella'.

Disease resistance	**50**
Flowering	**23**
Fragrance	**4**
Total rating	**77**

Large-flowered climber

Climbing to about 20 feet

Light pink

'New Dawn' can be considered a true star of the rose firmament. Its long-standing popularity is evident in many home landscapes. It is interesting to note that 'New Dawn' was the very first plant ever patented. The Plant Patent Act of 1930 granted the breeder or discoverer of a new plant variety the right to control its propagation and distribution for seventeen years. After that the variety is considered common property. US Plant Patent Number 1 was issued to New Jersey resident Henry Bosenberg on August 18, 1931, for 'New Dawn', which was described as having "champagne-colored blooms."

The disease resistance of New Dawn is legendary. Its blooms are the lightest pink and when the plant is in full bloom covering an arbor or tall fence, it can take your breath away. 'New Dawn' sported once again to a new variety called 'Awakening' which is also highlighted in this directory. 'New Dawn' was designated an Earth-Kind rose and declared in 1997 to be the World's Favorite Rose by the World Federation of Rose Societies, and inducted into the Rose Hall of Fame.

Introduced circa 1930 as a repeat blooming sport of 'Dr. W. Van Fleet'.

Companions | 'New Dawn' is a very vigorous climber with plenty of thorns, so be careful where you plant it and be sure to allow it plenty of room to grow. (I have seen unfortunate gardeners plant this rose next to a mailbox, which it quickly engulfed). I would accompany 'New Dawn' with other pastels like 'Blush Noisette', 'Eifelzauber', 'Peach Drift', 'Savannah', and 'Souvenir de Baden-Baden'. Whites and creams would work well too.

'Old Baylor'

Shrub

Tall, upright growth to 5–6 feet

White

'Old Baylor' has been a standout in the border ever since I have been growing it. I love the height it reaches and the covering of white blooms that it produces. The flowers are about 3 inches across and are usually borne in clusters. The white flowers are semi-double and open loosely flat to reveal a pretty presentation of bright yellow stamens; the effect is luminous. The petals have a somewhat scalloped edge to them, which lends a textured look to the bloom. The growth of 'Old Baylor' is strong and upright, and the foliage is a nice green that rarely shows signs of disease.

Introduced in 2003 by the Antique Rose Emporium.

Companions | 'Old Baylor' makes a fine specimen plant in a mixed border or a companion to other roses near the back of the border. The height of the plant and white color of the flowers combines well with most any other color. I like to plant roses that echo the color of the yellow stamens, such as 'Centennial', 'David Rockefeller's Golden Sparrow', 'Golden Gate', 'Postillion' and 'Sunny Sky'.

Disease resistance	53
Flowering	21
Fragrance	2
Total rating	76

Shrub

Bushy and spreading to about 3-5 feet

White

'Ole' was introduced along with 'Lena' as part of the Northern Accents series of roses bred for hardiness by Kathy Zuzek. Although it was bred for cold climates, gardeners in warmer regions can also plant this rose with confidence. 'Ole' has a bushy growth habit that makes it a stunning rose hedge. If you do not use it as a hedge, plant it in groups of three or five, or as a specimen in the mixed border. The soft cupped shape of the flower and the dark golden stamens offer a lot of flexibility to match this rose with other plants.

Introduced in 2007 by Bailey Nurseries; hybridized by Kathy Zuzek (BAIole).

Companions | I like to play off the flower form and the white color of 'Ole' with roses like 'Belinda's Dream', 'Easter Basket', 'Eliza', 'La Perla', and 'Wedding Bells'.

Disease resistance	54
Flowering	18
Fragrance	0
Total rating	72

'Oso Easy Cherry Pie'

Disease resistance	**60**
Flowering	**30**
Fragrance	**0**
Total rating	**90**

Shrub

Bushy and spreading to 3 feet

Medium red with a white eye

As implied by the moniker "oso," the plants sold under this name are said to be easy and carefree to grow. 'Oso Easy Cherry Pie' is one of my favorites from this collection. The simple red flowers open flat to reveal a delicious display of stamens that is set up by a white eye. The blooms are produced in clusters and seem to be never-ending. Other roses previously introduced in this series include 'Oso Easy Fragrant Spreader', 'Oso Easy Honey Bun', 'Oso Easy Mango Salsa', 'Oso Easy Paprika', 'Oso Easy Peachy Cream', and 'Oso Easy Strawberry Crush'. Newer varieties include 'Oso Easy Pink Cupcake', 'Oso Easy Double Red', 'Oso Easy Lemon Zest', and 'Oso Easy Italian Ice'.

Hybridized in 2006 by Meilland (MEIboulka).

Companions | This rose is stunning massed as a hedge or groundcover planting. As a single plant, it is suited to the middle or front of the border, or in a container, perhaps with some trailing foliage plants. Reds, yellows, blends, and whites are all very complementary with this shade of red. I suggest you plant it with 'Centennial', 'Darlow's Enigma', 'Innocencia Vigorosa', 'KOSMOS', and 'Lady Pamela Carol'.

Disease resistance	60
Flowering	27
Fragrance	0
Total rating	87

Miniature

Bushy and mounded to about 2–3 feet

Pink

This variety was hybridized by David Zlesak, one of my greatest teachers and a person who cares about disease resistance in roses more than almost anyone. 'Oso Happy Petit Pink' has proven to be a great garden performer and it provides continuous sprays of tiny (petit) pink flowers. The whole effect is indeed happy, and the care is minimal. 'Oso Happy Petit Pink' is a great rose for a container, a small garden, a rock garden, or along a border.

Introduced in 2011 by Proven Winners; hybridized in 2009 by David Zlesak (ZLE-marianneyoshida).

Companions | 'Oso Happy Petit Pink' is beautiful planted in numbers along a pathway or in a container. In the border, place it at the front where it can show its charm. The pink color is complementary with a lot of hues. I like to match it with other pinks, pink blends, yellows, and whites. Try this little rose with 'Innocencia Vigorosa', 'KOSMOS', 'Peach Drift', 'Savannah', and 'Stanwell Perpetual'.

'Out of Rosenheim'

Disease resistance	**60**
Flowering	**24**
Fragrance	**0**
Total rating	**84**

Floribunda

Bushy and upright to about 3 feet

Red

Red roses tend not be have the greatest disease resistance, but 'Out of Rosenheim' is a find, as it completely fits the bill when it comes to its ability to resist black spot. The dark green foliage is extremely healthy and makes a perfect backdrop for the magnificently romantic blooms. The deep red buds open to a perfectly cupped and ruffled flower that holds up well to the rain. 'Out of Rosenheim' grows strong from the first year and just keeps going strong. It was awarded a silver medal at the Baden-Baden Rose Trials in 2011 and a Certificate at the 2012 Royal National Rose Society Trials in England.

Introduced in 2010 by Kordes (KORmarkron).

Companions | 'Out of Rosenheim' is a great rose for the front of the border or even in a pot where you can enjoy the complexity of the flower form and the red color. Grow it with other bright colors or some softer peaches like 'Caramella', 'Eifelzauber', and 'Mother of Pearl'. Purple roses like 'Poseidon' accentuate the purple-blue base of the red color.

Shrub

Short and spreading to 2 feet

Apricot-peach

'Peach Drift' is an Energizer Bunny of a rose, pumping out blooms all season long. The individual flowers are small and open up to a soft, cupped form that reveals the stamens. 'Peach Drift' always seems to be smiling in my garden.

Introduced in 2008 by Conard-Pyle; hybridized by Meilland (MEIggili).

Companions | 'Peach Drift' would work well almost anywhere in the garden—in a massed planting, in a pot, along a path, or beside a patio. The buds have a yellow color and when they open up, the bright yellow stamens are revealed, so match with yellow roses such as 'David Rockefeller's Golden Sparrow', 'Julia Child', and 'Solero'. It also pairs nicely with pinks and whites such as 'Belinda's Dream' and 'KOSMOS'.

Disease resistance	60
Flowering	30
Fragrance	3
Total rating	**93**

'Peggy Martin Survivor'

Disease resistance	54
Flowering	25
Fragrance	4
Total rating	83

Large-flowered climber

Climbing to 15–18 feet

Pink

It's often the stories that go along with a particular rose that people seem to appreciate. This rose has become so famous that it even has its own website. I talk a lot about roses with tough genes, and this rose is a perfect example because it survived the flooded backyard of Peggy Martin in Plaquemines Parish, Louisiana, during Hurricane Katrina. After being inundated with 20 feet of water, the plant was alive and had lush green growth.

The stunning rose display is as gracious and charming as the person this rose is named after. The flowers are small, produced in clusters of frilly little pink blooms with shades of lighter pink to white. The canes are flexible and nearly thornless. In warmer climates, this rose can stretch to cover a fence in no time. 'Peggy Martin Survivor' deserves to be better known and grown for the wonderful qualities it can bring to the garden. In my garden it just continues to get better and better each year. Give it plenty of room to grow and watch it flourish with its generosity of color and fragrance in your garden.

Companions | This rose will happily take over a fence, pergola, or trellis. Make sure to give it some room so you can enjoy the full splendor of its flowers. Complement it with other "nostalgic" blooms and colors in front of it. Try 'Cinderella', 'Ducher', 'Pomponella', 'Quietness', 'Stanwell Perpetual', and 'Summer Memories'.

Disease resistance	50
Flowering	22
Fragrance	2
Total rating	74

Floribunda

Bushy, upright growth to 4–5 feet

Creamy white

One of the first qualities of this rose to catch my eye was the deep green, glossy foliage. Its growth is tall and somewhat lanky but it doesn't matter to me because high above the foliage appear these wonderfully frilly, old-fashioned flowers. Some of the petals are scalloped or notched—hence the name 'Petticoat'. The buds have a pleasing soft apricot color that is a perfect starting complement to the creamy white blooms, which themselves can carry a delicate apricot coloring near the center, particularly in the cooler weather. This is a lovely rose that has been recognized in international rose competitions and received the ADR Rose Award (2004) and a Certificate of Merit at the Australian National Rose Trials in 2007. This is another lovely, easy-to-grow rose that I would like to see in more gardens.

Introduced in 2004 by Kordes (KORgretaum).

Companions | With its tall growth, 'Petticoat' should be in the middle to the back of the border. Mix it with other creams, apricot-peaches, and reds. I suggest trying it with 'Cinderella', 'Eifelzauber', 'Garden of Roses', and 'Mother of Pearl', or with the lavender-purple of 'Thrive! Lavender'.

'Pink Bassino'

Disease resistance	55
Flowering	25
Fragrance	0
Total rating	80

Floribunda

Shorter, bushy growth to 20–24 inches

Lightest pink

I first planted this rose on the advice of a friend, and it has become one of those sleeper roses that just keeps going on and on. It is truly reliable in its bloom and health. The flowers are single, with a soft pink color that has some variability in it, almost like a subtle fading or streaking. The overall effect is very pleasing and traditional.

Introduced in 1991 by Kordes (KORbasren).

Companions | The shorter stature of 'Pink Bassino' means it is equally good for a mass planting, along walkways, or in the front of the border. The subtlety of the color is lovely paired with paler roses, especially creams and whites. I recommend pairing it with 'KOSMOS', 'Lion's Rose', and 'Polar Express', or with some darker pinks such as 'Belinda's Dream', 'Eliza', and 'Peach Drift'.

Disease resistance	59
Flowering	25
Fragrance	0
Total rating	**84**

Shrub

Low and spreading

Pink

Another in the Drift series of roses, 'Pink Drift' scores consistently high in trials year after year. Consistent and reliable are probably the best words to describe this rose: it is consistent in color, has constant bloom production, and exhibits superior disease resistance.

The blooms are produced abundantly above glossy, green foliage. They are small and open up fairly flat with a cheerful, frilly nature about them with some white tones toward the center. I have found that 'Pink Drift' is lower growing (with a much wider spread than height) than other Drift roses in the series. Planted in numbers, it will make a luscious green carpet covered with abundant pink blooms to be enjoyed on its own or in association with other roses and perennials.

Introduced in 2009 by Conard-Pyle; hybridized by Meilland (MEIjocos).

Companions | The spreading habit of 'Pink Drift' makes it a great candidate for constant color in the front of the border, along a pathway, or in a container. Match it with other shades of pink and white. Try it with 'Beverly', 'Carefree Beauty', 'Escimo', 'KOSMOS', 'Marie-Luise Marjan', and 'Polar Express'.

'Pink Martini'

Disease resistance	52
Flowering	25
Fragrance	0
Total rating	77

Floribunda

Bushy and very spreading to 5–6 feet wide and tall

Medium pink

This rose originally came into my garden as 'Pink Veranda'. The Veranda series of roses (previously Flower Circus) from Kordes is described by the breeder as "vigorous, yet compact disease-resistant rose varieties with 'Old World' or 'English Garden' flower shapes (including quartered, rosette, flat, pompon, and cupped blooms)." Their compact habit and repeat bloom make them ideal for patio containers and as compact roses for the garden." This is true for other Veranda roses from Kordes (such as 'Chica Veranda' and 'Brilliant Veranda'). However, it became clear that the vigorous growth and long stretching canes of 'Pink Veranda' did not fit the definition of compact so Kordes removed it from the collection and renamed it 'Pink Martini'. I love a martini—and here is a pink one!

'Pink Martini' grew so robustly for me that it was one of the varieties that I chose for the Northeast Earth-Kind trials. It has consistently scored very high and has grown to much bigger proportions than I ever could have imagined, consuming a fence in the meantime. The pink flowers are borne in profusion along the stems and above the nice green and extremely healthy foliage. The pink blooms are literally overflowing. That's my kind of martini!

Introduced in 2007 by Kordes as 'Pink Flower Circus' (KORfloci23); also sold as 'Moin Moin'.

Companions | Because of its vigorous growth habit, 'Pink Martini' makes a very colorful rose hedge. In the mixed border I would place it near the back as a large block of color. It might even do well in a very large urn where it has enough freedom to overflow and cascade. The color lends itself to a lot of wonderful combinations. For some orange-peach complements try 'Mutabilis' and 'Thrive! Copper', or pair it with other pinks like 'Cinderella', 'Dolomiti', and 'Garden Delight'.

China-Polyantha

Bushy, but stays to a compact 3 feet

Pink

I've planted this rose in masses along pathways and entrances where it can greet me with its constant display of pink (some say lilac) flowers that fade a little with age. The effect is of a multitude of soft pinks at any one time. The flowers are small and refined, with ruffled petals. A named Earth-Kind variety, this rose will give you lots of return for little effort.

Introduced in 1928 by George Lilley; found by Bill Welch in Caldwell, Texas, and reintroduced as 'Caldwell Pink'.

Companions | I have enjoyed planting 'Pink Pet' with many other roses that have old-fashioned flower forms. The color is very flexible to work with, and its compact size also makes it suitable for containers. Combine this rose with other Earth-Kind varieties like 'Belinda's Dream' or 'Ducher', or with roses such as 'Escimo', 'KOSMOS', and 'Lion's Rose'.

Disease resistance	58
Flowering	29
Fragrance	0
Total rating	87

'Plum Perfect'

Disease resistance	45
Flowering	30
Fragrance	0
Total rating	75

Floribunda

Bushy growth to 3–5 feet

Purple

It is hard to find good disease resistance in roses of this color. 'Poseidon' is a great example, and so is 'Plum Perfect'. It is a deeper plum shade compared to the lavender-gray of 'Poseidon'. The bud color is an even deeper shade with some pink tones.

The disease resistance is good and it has trialed well in hotter regions. The foliage is a shiny, dark green with distinctly wide leaves. This leaf characteristic helps to identify the plant even when it is not in bloom. The double flowers are set beautifully against the foliage and the overall effect is romantic and nostalgic.

Introduced in 2013 by Kordes (KORvodacom).

Companions | The growth habit of this rose makes it good for the middle of the border. In multiples of three, five, or seven, the effect will be exponentially greater. 'Plum Perfect' has a fun color to work with and I enjoy matching it with other purples, pinks, blends, apricot-oranges, and pastels. Try it with 'Alexandra Princesse de Luxembourg', 'Dark Desire', 'Sweet Fragrance', 'Thrive! Lavender', and 'Wedding Bells'.

Shrub

Bushy, upright growth to 4–5 feet

Creamy white

The name 'Polar Express' implies that something cool and white is heading our way. Lucky for us, instead of an arctic blast, we get lots of wonderfully shaped, semi-double blooms. Usually borne in clusters, the overall effect when this rose is in bloom is indeed that of a covering of snow that sometimes hides the vibrant, healthy foliage beneath.

Introduced in 2013 by Kordes.

Companions | This variety is appropriate for the middle of the border or in a spot in your garden where a cool color patch is needed. In a mass planting, it seems to offer a glimpse of winter to come. Complement 'Polar Express' with other pastels or combinations of white such as 'Blush Noisette', 'Carefree Delight', 'Mandarin Ice', 'Oso Easy Cherry Pie', 'Raspberry Kiss', and 'Ruby Ice'.

Disease resistance	56
Flowering	26
Fragrance	2
Total rating	84

'Pomponella'

Disease resistance	**60**
Flowering	**26**
Fragrance	**2**
Total rating	**88**

Floribunda

Upright, tall, and bushy to 6 feet

Pink

Several bushes of 'Pomponella' grouped together will give you literally thousands of rounded, globular pink blooms that are held high above the semi-glossy foliage beneath. If you have just one of these roses, you have to settle for hundreds of blooms instead of thousands. 'Pomponella' is in the Fairy Tale series of roses by Kordes, and like many other roses in this series, I have found it to be highly disease-resistant.

'Pomponella' has won many awards, including an ADR Rose Award (2006), a gold medal at the Gold Standard Rose Trials in England (2012), and a silver medal at The Hague Rose Trials (2010).

Introduced in 2005 by Kordes (KORpompan).

Companions | 'Pomponella' can be a specimen plant on its own, or it can be placed in groups with the same or other varieties. I have it planted with many other shades of pinks, salmons, and oranges. Try 'Alexandra Princesse de Luxembourg' to match it in height, or with 'Elegant Fairy Tale' and 'Quietness' for a lighter pink contrast. The bicolored 'Garden Delight' also complements 'Pomponella' beautifully.

Disease resistance	57
Flowering	26
Fragrance	9
Total rating	**92**

Floribunda

Upright, bushy growth to about 4 feet

Lavender

In Greek mythology Poseidon is the god of the sea, water, earthquakes, and horses and is called the Earth Shaker. The rose lives up to this mythological moniker, as lavender roses are rarely able to shake off disease pressure, but this is an exception. The growth on 'Poseidon' is very strong, consistent, and full. The flowers are borne in clusters, with a darker bud color like a cabernet wine that contrasts with the open lavender petals beautifully. The bloom is fully petaled and the edges of the petals are almost scalloped. This is a superb rose, able to tolerate a wide range of climates, and is an ADR Rose Award winner (2013).

Introduced in 2010 by Kordes (KORfriedhar); also sold as 'Novalis'.

Companions | 'Poseidon' has wonderfully deep pink-and-wine-colored buds that open to the loveliest lavender. Try complementing it with other roses that pick up the bud color such as 'Dee-Lish' and 'Laguna'. Any pastel rose would also blend beautifully, such as 'Cinderella', 'Eifelzauber', and 'Larissa'.

'Postillion'

Disease resistance	**55**
Flowering	**23**
Fragrance	**5**
Total rating	**83**

Shrub

Upright, tall growth to 5–6 feet

Yellow

As you may know by now, I have a fondness for yellow roses. (It is an emotional and personal connection with my mom. After all, roses connect with us on deep levels.) For years, it was frustrating trying to find yellow roses that didn't come with a bottle of fungicide. Now there seem to be quite a few with good disease resistance and 'Postillion' is one of the best. I planted it a few years ago as a very small rooted cutting. It took a couple of years for it to grow up, but now it has become one of my favorites, with strong, upright growth and foliage that is clean and healthy and wonderfully green. The blooms are held high above the foliage on sturdy stems that carry large clusters of flowers. This rose has won multiple awards, including the ADR Rose Award (1996), and silver medals at the Madrid Rose Trials (1995) and the Kortrijk Rose Trials (1994 and 2000), among others.

One of the stunning qualities of this rose is the multitude of colors. The orange buds open up to a bright yellow color that is as warming as sunlight. The flowers are semi-double, with about twenty petals. They open with an almost frilly effect and then become flatter to reveal bright orange stamens. The entire effect is that of a citrus tree, and it smells like one too—this rose doesn't seem to disappoint anyone who stops to smell the roses.

Introduced in 1998 by Kordes (KORtionza).

Coxmpanions | The taller stature of 'Postillion' lends itself to placement in the middle to back of the border, but make sure that you plant it where you can enjoy the fragrance. Other oranges, yellows, and whites would be great complements. I have paired it with 'Brilliant Veranda', 'Mandarin Ice', and 'Summer Sun', and with the white 'Darlow's Enigma' and 'Summer Memories'.

Disease resistance	60
Flowering	30
Fragrance	3
Total rating	93

Shrub

Dense and bushy to about 3 feet

Purple blend

When I first planted 'Purple Rain', the plants were so tiny I was sure there was no way they were going to survive, let alone grow up. But not only have they grown profusely into incredibly strong plants, I now consider this one of the best shrub roses I have ever grown. 'Purple Rain' consistently scores the highest scores year after year in rose trials, and has won gold medals in international rose trials. In fact our evaluations were delayed the year Hurricane Irene came up the East Coast. It rained more than 20 inches in just a few days. When we were finally able to do our evaluations after the storm, many of the roses had been hit hard. Not 'Purple Rain'—it shrugged off that hurricane, as its name might have suggested.

The foliage is exceptionally beautiful, giving the plant value in the landscape even when not in bloom. Not only is it the deepest of greens and healthy (with not a speck of black spot) but the leaves have a very fine, almost fernlike texture. The plant sends up longer canes that can be somewhat arching. These canes carry rich clusters of the small, complex, and frilly purple flowers above the foliage in a wonderful, charming floral presentation that is continuous throughout the growing season.

In my Maine garden, this is the rose I chose to plant along a long allée to the ocean. At a distance, the deep green foliage almost looks like boxwood, if boxwood had pretty purple flowers.

Introduced in 2009 by Kordes (KORpurlig).

Companions | Put compact 'Purple Rain' in a pot, in a mass planting as a rose hedge, or as a color blast in the mixed border. Try contrasting it with orange and golden-yellow roses like 'Brothers Grimm', 'Eifelzauber', 'Mandarin Ice', 'Michel Bras', and 'Mother of Pearl'.

'Quietness'

Disease resistance 48
Flowering 23
Fragrance 6
Total rating 77

Shrub

Upright, and tall to 5 feet

Light pink

'Quietness' is one of my all-time favorite roses, and I find the fragrance divine. As you would expect of a rose hybridized by Professor Buck, it is very hardy. The flowers are single or borne in clusters and the form is classic, with pointed buds. I have experienced some years where black spot can be found on this rose but its tough genes seem to pull it through, and all of the other wonderful qualities of this rose help me look past some disease issues.

Introduced by 2003 by Roses Unlimited; hybridized by Griffith J. Buck.

Companions | 'Quietness' can be a specimen placed where you can treat your senses to the exquisite fragrance. I have planted it with other tall pink roses like 'Eliza' and 'Pomponella'. Any white or cream rose would be a beautiful companion, such as 'Summer Memories'.

Disease resistance	59
Flowering	23
Fragrance	0
Total rating	82

Floribunda × *Hulthemia persica*

Bushy growth to 3 feet

Light pink with a blotched center

'Raspberry Kiss' emerged from efforts to hybridize roses with their wild relative *Hulthemia persica* over the past few decades. The goal is to bring into roses the genes that produce a desirable dark blotch found on the petals of *H. persica*. Some hybrids of *H. persica* have been marketed as the Eyeconic rose series, but although they are captivating and unique flowers, I have not found them to be very disease resistant. As usual, if you breed a rose for one particular attribute, you can easily lose others.

With this said, I was given 'Raspberry Kiss' to try as one of the first of the *Hulthemia* crosses meant to be disease resistant. In my humid New York garden, 'Raspberry Kiss' was completely disease free. The clustered single flower of this plant has a pleasing pink hue that is a perfect background for the dark "eye," which is itself a perfect background for the subtle yet wonderful stamen display sitting proudly above. The buds are a dark pink, so the total color effect is anywhere from light cream to pink to darker pink to the darkest pink in the eye. It was awarded numerous prizes in rose trials, including at The Hague Rose Trials (2009), in France at the Bagatelle Rose Trials (2011), and the Orléans Rose Trials (Environmental Prize, 2012).

Introduced in 2013 by Warner's Roses (CHEwsunsign); also sold as 'Eyes on Me'.

Companions | The wonderful pink of this rose gives you a lot of flexibility when matching with other colors. 'Raspberry Kiss' is such a conversation piece that it would do well in a container, in a large planting along a path, or in the front of the border. Mix it with other pinks and whites to highlight the variations of hues in this bloom. I like to grow it alongside 'Flamenco Rosita', 'KOSMOS', 'Macy's Pride', 'Ruby Ice', and 'Summer Memories'.

'Republic of Texas'

Disease resistance	50
Flowering	23
Fragrance	2
Total rating	75

Shrub

Compact, bushy, and somewhat spreading to about 3 feet

Light yellow

'Republic of Texas' is one of the best from the Texas Pioneer collection. It is also one from the collection that has consistently scored high in rose trials over the years. Surprisingly, the individual bloom isn't that much to look at: small, flat, with about fifteen petals. But these flowers are produced all over the plant in the most charming way. Every time I walk by 'Republic of Texas' I pause because of the sight of all these small, slightly ruffled blooms glowing in the sunshine. The deep green foliage is a nice contrast for this wealth of warmth and friendliness.

Introduced in 2000 by the Antique Rose Emporium.

Companions | 'Republic of Texas' would be a great variety for a container, and it also works well planted in a mass along a pathway or in the front of the border. I grow it combined with pastels like 'Easter Basket', 'Fortuna Vigorosa', 'La Perla', 'Mutabilis', and 'Peach Drift'.

Disease resistance	50
Flowering	19
Fragrance	4
Total rating	73

Noisette

Climbing to 15–20 feet

Yellow with darker shading

'Rêve d'Or' is best suited to warmer conditions, so if you garden further north, this rose is not going to be happy. Like 'Climbing Pinkie', 'Rêve d'Or' has flexible canes that are thin, pliable, and easily manipulated so that you can easily train it onto any garden structure. Its overall effect in the landscape is outstanding. The lack of thorns is also a plus if people frequently walk under this rose as they pass through an arbor or trellis. The blooms are a yellow color with darker shades of apricot or copper. I've noticed that in cooler temperatures, there is even more color in the blooms. Like many Noisettes, each individual flower has a softly cupped shape. 'Rêve d'Or' can be left alone; in fact, it prefers not to be pruned. It has been designated as an Earth-Kind rose.

Introduced in 1869 by Ducher.

Companions | Grow 'Rêve d'Or' over a fence, light post, arbor, or pergola, and accompany it with other pastels, apricot tones, and whites. Possible companions include 'Michel Bras', 'Sweet Fragrance', 'Savannah', 'Thrive! Lavender', and 'Thanksgiving Rose'.

'Roemer's Hip Happy'

Disease resistance	**53**
Flowering	**23**
Fragrance	**1**
Total rating	**77**

Shrub

Bushy, compact growth to 3–4 feet

Pink

Ferdinand von Roemer was a German geologist who visited the United States in 1845 to study the geology of Texas and other southern states. I suspect he was a good-natured man, because 'Roemer's Hip Happy' certainly has a pleasantness about it that brings a smile to my face. The roses are small in size and simple in form, and there is a charm to them that is truly captivating.

Introduced in 2004 by the Antique Rose Emporium.

Companions | 'Roemer's Hip Happy' is beautiful in a mass planting, but it also looks great in a pot, or planted along a pathway or at the front of a mixed border. The lighter creamy white center of the bloom and the fine presentation of stamens can be complemented with roses like 'Ducher', 'Elegant Fairy Tale', 'Innocencia Vigorosa', 'KOSMOS', and 'Solero'.

Disease resistance	60
Flowering	27
Fragrance	3
Total rating	90

Large-flowered climber

Tall and sturdy, climbing to 15 feet

Salmon pink

Without hesitation I will state that this is one of the best climbers I have ever grown. 'Rosanna' is a very strong grower, sending up large and sturdy canes that are covered in deep green, glossy foliage. The canes need to be strong to support the flowers, which are cupped blooms up to 5 inches wide. When the plant is covered with a profusion of those voluptuous salmon-pink blooms, it is enough to stop you in your tracks. 'Rosanna' has consistently rated at the top of my trials.

Introduced in 2002 by Kordes-Newflora (KORhokhel).

Companions | Give 'Rosanna' some room to shine and indeed it will. The color is unusual for a climber and I have had great fun placing other pastels and vivid colors in front of it. Try some lighter colors like 'Cinderella', 'Dolomiti', 'Larissa', and 'Lion's Rose', or roses with hotter tones such as yellow 'Postillion' with its complementary orange buds.

'Rosenstadt Freising'

Disease resistance	**58**
Flowering	**24**
Fragrance	**0**
Total rating	**82**

Shrub

Bushy habit spreading to 3–4 feet

Cream with hot pink edges

This is a rose that you don't see often enough in home gardens, and that's something I would like to change. 'Rosenstadt Freising'—which translates to "rose city of Freising"—may be a challenge to pronounce but this rose could easily be called 'Wow' for the colors in its buds. They are a creamy white with a glowing apricot-pink center. The petals have a tinge of pink dotted like lipstick around their edges, with a bright pinkish-red mark in the center.

They almost seem to be hand-painted and they are certainly photogenic. Surprisingly, the flower opens up to a solid color that I would call dusty pink or dusty apricot-pink.

The foliage is extremely healthy and the plant sends out long, arching canes in a spreading habit. At the end of each of these canes are found these remarkable kaleidoscopes of color. This rose has done well in international competitions where it has won silver medals at the Rose Trials in Kortrijk, Monza, and Rome (all 2002), and a gold medal in Geneva (2003), among others.

Introduced in 2006 by Kordes (KORcoptru).

Companions | With its short, spreading habit, 'Rosenstadt Freising' can be placed at the front of the border or along a pathway. Placed in a container, it would be a conversation piece. I have matched roses in shades of pink and red with it, such as 'Beverly', 'Eliza', 'Fiji', 'Flamenco Rosita', and 'Ruby Ice'.

Miniature

Small and bushy to 2 feet

Bright pink

As long as I have been growing this rose, it just keeps getting better and better—with little effort on my part. 'Roxy' is a compact plant that sends out panicles of bloom above its highly glossy, disease-resistant foliage. Sometimes the flower clusters are so abundant that there is no visible greenery. If you like a blast of pink (some say purple) in your garden, 'Roxy' is a great choice. It has consistently scored high in our monthly and yearly trials, and received an ADR Rose Award (2008) and the Frederick Law Olmsted Award for Best Groundcover at the Biltmore International Rose Trials in 2013. This is truly a rose to plant, forget about, and then sit back and enjoy year after year.

Introduced in 2007 by Kordes (KORsineo).

Companions | 'Roxy' is compact enough for a pot, or for the front of the border. The strong color can hold its own with other purple roses such as 'Poseidon' and 'Thrive! Lavender' or orange varieties like 'Garden Delight'. Softer pinks such as 'Cinderella' and 'Larissa' will give a nice contrast of flower forms and plant heights.

Disease resistance	60
Flowering	30
Fragrance	0
Total rating	90

'Ruby Ice'

Disease resistance	58
Flowering	30
Fragrance	0
Total rating	88

Floribunda

Bushy, upright growth to 3–4 feet

Ruby red with a silvery white reverse

'Mandarin Ice' is a brilliant deep orange and 'Ruby Ice' is a deep ruby red, but both have a silvery white reverse that makes quite a contrast to the deep, saturated colors of these roses. 'Ruby Ice' is a 2009 ADR Rose Award winner. It grows into a very full shrub with beautiful, lustrous green foliage. The flowers are produced in large clusters that keep coming through the growing season. The buds are beautifully presented and very tightly held. As the red bud unfurls, it reveals more of its reverse color. The petals have a scalloped edge. The whole effect is like a highly decorated wedding cake.

Introduced in 2011 by Kordes (KORburox); also called 'Schöne Koblenzerin'.

Companions | 'Ruby Ice' is beautiful in mass plantings, in the middle of the mixed border, or shown off in a container. The highly saturated red color and silvery reverse suggest some possible color schemes. Try some other reds, pinks, and whites like 'Innocencia Vigorosa', 'Macy's Pride', 'Oso Easy Cherry Pie', 'Out of Rosenheim', and 'Polar Express'.

Disease resistance	60
Flowering	25
Fragrance	9
Total rating	94

Hybrid Tea

Bushy, upright growth to 4 feet

Apricot-pink

'Savannah' came to my knowledge from a grower friend in Texas who couldn't say enough about it. Most parts of Texas have high temperatures and frequently experience drought conditions, so it takes a tough rose to do well there, and 'Savannah' is cer-tainly tough. The blooms are a perfect old-fashioned quartered type, but they look as fresh as tomorrow. This is a little different flower form than the classic Hybrid Tea, but like most roses in this class, there is typically one bloom per stem, nicely set off by the very disease resistant foliage. In cooler temperatures, I have found that the flower color becomes more saturated.

Introduced in 2013 by Kordes (KORvioros).

Companions | Match 'Savannah' to roses in shades of pastels, lavenders, and pinks. Try 'Cinderella', 'Lion's Rose', 'Mother of Pearl', and 'Souvenir de Baden-Baden'. The deeper apricot color as the bud opens would match well with other oranges like 'Michel Bras'.

'Sea Foam'

Disease resistance	49
Flowering	20
Fragrance	2
Total rating	**71**

Shrub

Bushy, rambling habit to 5–8 feet

White

This conversation about roses without chemicals would not be complete if it didn't include roses like 'The Fairy', 'La Marne', and 'Sea Foam'. These are all reliable disease-resistant roses. As the name implies, 'Sea Foam' resembles a rambling, crashing wave edged with white foam. The individual blooms are quite small; however, they are produced in great clusters that cover the plant. With proven toughness in the landscape—and an Earth-Kind designation—this is a rose that can add some drama to a fence or when cascading down a bank. It can even be grown as a standard. The parentage of 'Sea Foam' includes climbing roses. Too often I have seen it grown in an area where the gardener tried to keep it neat and tidy. Just let it sprawl and enjoy the crashing surf.

Introduced in 1963 by Conard-Pyle; hybridized by Ernest W. Schwartz.

Companions | Plant 'Sea Foam' in an area where it can stretch out and be dramatic. It can also be trained as a climber or arched over a fence. Its white flowers are a good foil for a wide variety of color matches in the garden. Try it with 'Alexandra Princesse de Luxembourg', 'Alister Stella Gray', 'Blush Noisette', 'David Rockefeller's Golden Sparrow', and 'Morning Magic'.

Disease resistance	60
Flowering	23
Fragrance	0
Total rating	83

Floribunda

Medium, lax, and spreading to about 3 feet

Pink

Maybe because of its lax habit and the fact that the flowers are not held high above the plant, 'Sister's Fairy Tale' isn't sold as much as it should be. As a home gardener who loves to grow healthy, disease-free plants with pretty flowers, I consider this one of the best. I have planted several other varieties around or near 'Sister's Fairy Tale' and they all seem to pale in comparison with its lush glossy green foliage and spreading habit. I don't think I have ever seen any disease on this plant in the years that I have grown it. It is a beautiful, spreading, glossy green plant. The blooms are fully double, very old-fashioned, and usually produced in clusters. It's true that the flowers are somewhat pendulous, but from a distance this variety has wonderful pink accents from light to dark.

In 2004, 'Sister's Fairy Tale' won a Certificate of Merit in the Bagatelle Rose Trials and a bronze medal in the Australian National Rose Trials.

Introduced in 2001 by Kordes (KORgrasotra); also sold as 'Home and Garden'.

Companions | Mass 'Sister's Fairy Tale' as a hedge or along a walkway or border, making sure to give it plenty of room for its spreading habit. The pleasing pink color is easy to blend with nearly anything. Any pinks or pastels will complement this rose, as will many blends, so try 'Larissa', 'Mother of Pearl', 'Mutabilis', 'Peach Drift', and 'Quietness'.

'Solero'

Disease resistance	52
Flowering	24
Fragrance	2
Total rating	78

Floribunda

Shorter, bushy, and spreading to about 3 feet

Yellow

'Solero' is another yellow rose with good disease resistance. The extremely healthy foliage is a glossy green stand-out among other roses. The flowers seem to be produced in a constant flow. They bloom in clusters, opening up to a fairly flat form showing an abundance of petals, sometimes with a button eye, that are almost old-fashioned in appearance. Like 'Belinda's Dream', this variety doesn't drop its petals cleanly. 'Solero' is fairly short and somewhat spreading, making it a great plant and color swatch to place at the front of the border.

I'm always pleased to see a yellow rose like 'Solero' receive international recognition for its quality. It has been given numerous awards, including the ADR Rose Award (2009) and silver medals at the Rome Rose Trials (2009) and at the Kortrijk Rose Trials (2011), among others.

Introduced in 2009 by Kordes (KORgeleflo).

Companions | 'Solero' is a shorter grower, so plant it in the front of the border or alongside a patio or pathway. It would also be great in a pot. Try mixing it with some oranges, pinks, or lighter colors to really complement the soft charming yellow of its blooms. I recommend 'Easter Basket', 'Garden Delight', 'Marie-Luise Marjan', 'Pink Drift', and 'Topolina'.

Disease resistance	**56**
Flowering	**22**
Fragrance	**6**
Total rating	**84**

Hybrid Tea

Upright, medium growth to about 3–4 feet

Cream with pink edges

There are times when the nomenclature of roses can be confusing. This rose is a case in point. It was originally introduced in Germany as 'Souvenir de Baden-Baden', but it was renamed for the US market as 'Pink Enchantment'. You can find it sold under both names. Frankly, you could call it 'Ralph' and I would still grow it in my garden!

This 2010 ADR award-winning rose (among many other accolades) is truly enchanting as the name implies. As a typical Hybrid Tea, it has flowers borne singly, one per long stem. In bud form, they are a soft cream with just the slightest hint they might be concealing another color. As the flower opens, the bloom retains its creaminess but reveals some gorgeous peachy pink tints and edges—quite a nice combination. The blooms are very large and have a lot of substance, with scalloped edges. The foliage is healthy, deep green, and glossy and a perfect background for the lighter shades of the blooms. The long stems make this a great cut flower for arrangements, with a fragrance that is often described as complex with powdery and fruity notes. This rose is indeed an enchanted souvenir for any garden.

Among the other awards this rose has received are The Hague Rose Trials Fragrance Award (2013), the Hradec Rose Trials First Prize (2012), the Dublin Rose Trials winner (2011), and the Nagasaki Rose Trials gold medal (2010).

Introduced in 2008 by Kordes (KORsouba); also sold as 'Pink Enchantment'.

Companions | The growth habit of 'Souvenir de Baden-Baden' lends itself to be more of a specimen near the middle to front of the border. Make sure you plant it as a featured bush on your stop-and-smell-the-roses route through the garden. Pastels, creams, and whites are all complementary, so it can be partnered with many other roses, including 'Beverly', 'Cinderella', 'Dolomiti', 'Easter Basket', 'Eliza', 'Larissa', 'Pink Pet', 'Pomponella', 'Quietness', 'Rosenstadt Freising', and 'Thérèse Bugnet'.

'Spice'

Disease resistance	50
Flowering	20
Fragrance	7
Total rating	77

China

Bushy, upright to about 3–5 feet

Blush pink to white

This rose is something of a mystery, believed to be 'Hume's Blush Tea-Scented China' which is an ancestor of countless modern roses. 'Spice' is a designated Earth-Kind variety and it is indeed tough and long lasting. The bush produces wave after wave of blush-pink, double blossoms that are lighter pink in the cooler weather and mostly white in the heat of the summer. These blooms have a fragrance that has been described as peppery, which probably gave rise to the name. 'Spice' is a shrubby grower, gaining more height and width in warm regions, but it's more likely to shine in an informal border complemented with companions of other colors.

Companions | This variety is appropriate for the middle to the front of the border or in a container—or anywhere you can enjoy its mysterious fragrance. For rose companions, I like to keep with roses that have a similar style and color palette, such as 'Belinda's Dream', 'Cubana', 'Darlow's Enigma', 'Ducher', 'Easter Basket', 'KOSMOS', 'Marie Daly', and 'Quietness'.

'Stanwell Perpetual'

Hybrid Spinosissima

Bushy growth to 4–5 feet

Light pink

The Spinosissima class of roses gives us some exceptionally beautiful plants, and they produce some very charming roses. This is a very prickly plant with great foliage and texture. The blooms are double, somewhat flat in their form, and often sporting a button eye—a characteristic of some very full-petaled old roses where the center of the blooms is incurved and resembles a button. As a bonus, these flowers are free flowing and wonderfully fragrant (even I have to admit that).

Introduced in 2004 by the Antique Rose Emporium; hybridized in the UK circa 1830s by C. Brown and Lee.

Companions | I use this striking rose in front of the border, or in the middle where its gray-green foliage can provide an unusual foil for flowering plants with lighter pastel colors. I have planted it with yellows, pastels, and whites to great delight. Some suggestions include 'Ducher', 'Easter Basket', 'Souvenir de Baden-Baden', 'Thérèse Bugnet', and 'Wedding Bells'.

Disease resistance	58
Flowering	19
Fragrance	5
Total rating	82

'Summer Memories'

Disease resistance	**59**
Flowering	**25**
Fragrance	**2**
Total rating	**86**

Shrub

Arching and spreading to 4 feet or more

White

This is a variety with old-style charm, fragrance, and healthy foliage. 'Summer Memories' never seems to be out of bloom and is constantly pushing new growth from each leaf axil as it grows. The shrub fills out very quickly, making one-year-old plants in the garden seem as though they have been there for years. The health and vigor of the foliage is outstanding, the dark green glossiness of the leaves providing a nice background to the abundance of creamy to pure-white blooms, which can occur singly or in a small cluster of three to five. This makes a good substitute for 'Iceberg' which can be prone to black spot in high disease pressure regions. 'Summer Memories' is a great addition to any rose garden, perennial border, or white garden.

Introduced in 2004 by Kordes (KORuteli).

Companions | The color of this rose allows it to be really flexible in the garden. As a solid white it will pair well with any other solid color, or with other flowers that have a white blend, such as 'Alister Stella Gray' (light yellow fading to white), 'Carefree Delight' (pink with white eye), 'David Rockefeller's Golden Sparrow' (light yellow fading to white), and 'Oso Easy Cherry Pie' (red with white eye). Pinks like 'Eliza' are also a good match.

Disease resistance	52
Flowering	28
Fragrance	10
Total rating	90

Floribunda

Bushy, upright habit to 4 feet

Dark pink

This rose belongs to the Parfuma collection by Kordes Roses and like many of their modern releases it combines disease resistance with intensely fragrant, romantic blooms. 'Summer Romance' is extremely vigorous, growing very bushy and covering itself with highly scented blooms. The buds are a dark pink and open up to reveal a quartered cup-shaped bloom—much like a rose from centuries past. The dark green foliage is glossy and very disease-resistant. The fragrance is described as extraordinary, startling the senses with notes of sharp citron, berries, and apples. This rose represents everything you could desire in a rose for the chemical-free garden.

Introduced in 2014 by Kordes (KORtekcho); also known as 'Rosengräfin Marie Henriette'.

Companions | The wonderful pink of this rose will give a lot of flexibility in color matching. Keep 'Summer Romance' toward the edge of the border where you can fully enjoy the fragrance. I like to mix it with other pinks and combinations, such as 'Dolomiti', 'Peach Drift', 'Savannah', 'Souvenir de Baden-Baden', and 'Wedding Bells'. Creams and whites would work well too.

'Summer Sun'

Disease resistance	**60**
Flowering	**25**
Fragrance	**1**
Total rating	**86**

Floribunda

Bushy, upright growth to about 4 feet

Orange blend

Although 'Summer Sun' is classified as a Floribunda, I think of it more as a Shrub rose. It is an award-winning (ADR Rose Award 2007, among others), vigorous grower with extremely lush and healthy foliage. I don't think I have ever seen any leaf spot on this plant in my garden. This verdant foliage is a perfect backdrop for the very colorful blooms, which open up to a salmon-orange color with a peachy golden reverse. The individual flowers are 3–4 inches, and are usually borne singly or in clusters. The blooms have a very full and blousy look to them—this is not a dainty rose. The whole effect is like a large colorful bouquet that resembles the setting sun. Although the color of 'Summer Sun' is not as intense as that of 'Brothers Grimm', the two varieties show the same luxuriant characteristics.

Introduced in 2010 by Kordes (KORfocgri); also sold as 'Sommersonne'.

Companions | 'Summer Sun' can grow very bushy and is spectacular in a mass planting. It is also stunning placed in the middle of the border and mixed with complementary sunset colors, such as 'Caramella', 'Coral Drift', 'Mutabilis', 'Postillion', and 'Thrive! Lavender'.

Disease resistance	55
Flowering	22
Fragrance	3
Total rating	80

Hybrid Tea

Bushy, upright growth to about 4 feet

Yellow

I am thrilled to highlight another yellow rose here that has superior disease resistance. 'Sunny Sky' is very vigorous in its growth and very healthy in its glossy green leaves. Even though this is a Hybrid Tea rose, the flowers are sometimes produced in wonderfully large clusters set on strong stems high above the foliage. The buds have a darker orange hue and open up with luminous charisma to fill a garden. The name 'Sunny Sky' is so appropriate as these classically high-centered flowers do seem to be reaching for the sky. If you like yellows as much as I do, add this variety to your garden.

'Sunny Sky' has won international awards for its quality, including gold medals in Monza (2010) and La Tacita (2012). In 2012 it was also declared the Best Large-flowered Rose/Best Hybrid Tea in Belfast.

Introduced in 2009 by Kordes (KORaruli).

Companions | 'Sunny Sky' is a medium grower so would do well in the middle of the border or more near the front where you can enjoy the cheerfulness of the colors. The orange buds and the fading of the outer petals as they open up really complement the honey-yellow color in the center of each flower. Pair it with some other oranges or lighter colors to really accentuate this effect. Try 'Brothers Grimm', 'Caramella', 'Garden Delight', 'Innocencia Vigorosa', 'Michel Bras', and 'Old Baylor'.

'Super Hero'

Disease resistance	54
Flowering	22
Fragrance	0
Total rating	76

Floribunda

**Medium, upright growth
to about 4 feet**

Red

'Super Hero' is part of the Easy Elegance collection of roses and I've had good success growing many of these. This variety has very healthy, deep green glossy foliage and vivid red blooms. The flowers are on the smaller side, about 2 inches across, each with around thirty petals. They somewhat resemble a Hybrid Tea form and unfurl to a nice cupped shaped. There is a tinge of some pink in the blooms to highlight as they finish. You don't have to be a superhero to grow this red rose!

Introduced in 2007 by Bailey Nurseries (BAIsuhe).

Companions | My experience with 'Super Hero' is that it can get tall by the end of the season, so plant it in the middle to the back of the border and highlight it with other strong, saturated colors, pinks, blends, or creams. Good rose companions include 'Caramella', 'Garden Delight', 'Mutabilis', and 'Tequila'.

Disease resistance	53
Flowering	21
Fragrance	5
Total rating	79

Grandiflora

Smaller, upright growth to about 3 feet

Apricot blend

'Sweet Fragrance' is another of the Easy Elegance collection. This is a nice variety that delivers on many levels. The foliage and health of plants in this collection are a given. 'Sweet Fragrance' has scored consistently high in my trials and I have rarely seen any leaf spot on it. It has a beautiful color palette, with buds that are nicely tinted with apricot and unfurl to reveal even more colors of coral to yellow. The flowers are borne in clusters, each bloom about 3–4 inches across with twenty to thirty petals, opening fairly flat with an informal look. The finish color is more a peachy pink.

Introduced in 2007 by Bailey Nurseries (BAInce); also sold as 'Easy Elegance'.

Companions | 'Sweet Fragrance' is fairly small, so put it near the front of the border where you can enjoy the color play. I could imagine it in a mass planting as well. Complement the colors of the 'Sweet Fragrance' color blend with other solid colors like 'Brilliant Veranda' and 'Eifelzauber', or blends such as 'Caramella', 'Thrive! Copper', and 'Thrive! Lavender'.

'Sweet Jane'

Floribunda

Bushy, spreading habit to 3–4 feet

Golden apricot

A sweet name for a sweet rose! This is a beautiful Floribunda rose with gracious blooms that have a wonderful old-fashioned look about them, almost like tissue paper roses. They open in a cupped form to reveal shorter petals in an almost quartered form and they seem to be smiling. I believe the color is really special, and in my years of working in public gardens, this color is probably one that stops visitors most often. 'Sweet Jane' has trialed very well in the heat and humidity, proving its toughness in the hardest of conditions.

Introduced in 2014 by Kordes-Newflora (KORflusamea).

Companions | The bushy habit of 'Sweet Jane' places it nicely in the middle to front of the border. I like it in a place where the bloom color is really well displayed. The golden-apricot hues offer some great combinations. Try it with 'Brothers Grimm', 'Caramella', 'Garden Delight', 'Mutabilis', and 'Out of Rosenheim'.

Disease resistance	45
Flowering	25
Fragrance	0
Total rating	70

Disease resistance	58
Flowering	20
Fragrance	0
Total rating	78

Floribunda

Short, bushy, and spreading to 2–3 feet

Deep pink

'Sweet Vigorosa' is part of the Vigorosa series of roses by Kordes. In my garden roses from this series have a compact habit and provide a reliable low patch of constant color. 'Sweet Vigorosa' is one of the strongest I have seen in the series (along with 'Innocencia Vigorosa'). The growth habit is strong, with little to no black spot to be found on the glossy foliage.

Introduced in 2001 by Kordes (KORdatura).

Companions | The compact habit of this plant is perfect for pots, mass plantings, or near the front of the border. The showy yellow stamens are a nice color to enhance with other yellows or golds. Try it with 'Centennial', 'David Rockefeller's Golden Sparrow', 'Solero', 'Thrive! Lemon', and 'Winter Sun'.

'Tequila'

Disease resistance 56
Flowering 20
Fragrance 0
Total rating 76

Floribunda

Medium bushy to about 4–5 feet

Yellow-apricot blend

With 'Bonica' in its parentage (a classic powerhouse of a shrub), 'Tequila' demonstrates the characteristics of a strong grower with seemingly constant color and reliability. I have found that the color of this rose can be affected by temperature, with cooler temperatures giving it more of an apricot color and hotter temperatures making the color fade to soft yellow and faded yellow. The overall effect is pleasing and soft. 'Tequila' is a sturdy grower with wonderfully green foliage. The flowers are about 2–3 inches and are generally produced in clusters. 'Tequila' has another color sport, 'Tequila Gold'. It's like a shelf of fine bottles of 'Tequila' to choose from—indulge, and call me in the morning!

Introduced in 2006 by Meilland (MEIpom-olo).

Companions | Have fun combining this rose with other oranges like 'Brilliant Veranda', 'Brothers Grimm', and 'Coral Drift'. The yellow stamens against the white flowers of 'Innocencia Vigorosa' and 'Macy's Pride' also complement the yellow variations in 'Tequila'.

Disease resistance	59
Flowering	18
Fragrance	3
Total rating	**80**

Hybrid Tea

Bushy and upright to 4–5 feet

Apricot

'Thanksgiving Rose' has one of those colors that always seems to be a crowd favorite, and the name is apt because the tones in this rose suggest changing of seasons from summer to fall. This rose has a wonderful, robust growth habit with deep green foliage that is extremely healthy. The foliage is a perfect setting for the buffet of flowers, borne one per stem and apricot filled with darker tones—much like sweet potatoes. I have seen them in clusters a few times. The thick, substantive petals don't seem to unfurl like the classic Hybrid Teas but stay more closed, never really revealing the center of the flowers. Admirers comment on the fragrance, which just completes this feast of a rose.

'Thanksgiving Rose' was awarded the Baden b. Wien Rose Trials First Prize (2009), the Lyon Rose Trials Grande Rose du Siècle (2009), and the Buenos Aires Rose Trials gold medal (2008).

Introduced in 2008 by Kordes (KORwawibe).

Companions | 'Thanksgiving Rose' can be planted in groupings of three or five. Complement it in a mixed border with colors that work well with its unusual apricot tones, such as 'Coral Drift', 'Mutabilis', 'Postillion', 'Savannah', and 'Tequila'.

'The Fairy'

Disease resistance	50
Flowering	23
Fragrance	1
Total rating	74

This is a long-established and easy-to-grow rose. 'The Fairy' is a designated Earth-Kind variety, indicating that it is one of the tough garden performers that is meant to last. The growth habit of 'The Fairy' is short and spreading, sometimes even arching. The overall growth habit is very informal and subtle. The individual blooms are about a half-inch across, blooming in clusters that seem to be in constant production. I have found this rose to be excellent for resistance to black spot and other diseases.

Introduced in 1941 by Conard-Pyle; hybridized in 1932 by Bentall.

Companions | The soft effect of this plant is useful in so many ways in the landscape. You can mass it for a nice hedge or use it individually in a pot or near the front of the border. The smaller pink blooms in clusters are a fine complement to larger-blooming roses. Its pink color tends to fade to lighter shades and so pairs well with roses such as 'KOSMOS' and 'Lion's Rose'. You can also match it with taller 'Blush Noisette' or 'Wedding Bells'.

Disease resistance	55
Flowering	19
Fragrance	9
Total rating	83

Hybrid Rugosa

Tall, upright growth to 6 feet

Medium pink

One could certainly write an entire book about Hybrid Rugosa roses for the chemical-free garden. 'Thérèse Bugnet' is one of my all-time favorite Hybrid Rugosas, so even though it not one of the latest hybridized roses, it has proven itself in the landscape time and time again. This rose deserves to be grown in more gardens for a multitude of reasons.

Given the inherent disease resistance and hardiness thanks to its genes, the foliage is a beautiful and even green. The buds are a darker pink color, almost red, and open up to a medium-pink bloom. The flowers are 3–4 inches across and have a loose quality about them although they do have a classic nostalgic form and tend to have a button eye. The fragrance of 'Thérèse Bugnet' can fill the garden. When the weather turns cold, the stems turn a wonderful red color, giving good winter interest.

Companions | 'Thérèse Bugnet' has such a presence it could be grown as a specimen plant, in a mass as a wonderful rose hedge with winter interest, or in the back of the border. The pink color is easy to work with, so try it with some other shades of pinks and cream like 'Alexandra Princesse de Luxembourg', 'Blush Noisette', 'Dolomiti', 'Souvenir de Baden-Baden', and 'Summer Memories'.

'Thrive!'

Shrub

Bushy growth to 4 feet

Red

The Thrive! series are among the best examples of recent hybridization efforts toward disease resistance in roses. Planted in mass or as a hedge, any of the Thrive! roses makes an excellent bank of color. The true red color of 'Thrive!' is bright and hard to miss in the garden. The loose petals (about seven to nine per bloom) open flat to reveal a nice display of yellow stamens. The blooms are produced in generous clusters offering continued color in the landscape.

Introduced in 2011 by Conard-Pyle; hybridized by Jim Sproul (SPRothrive).

Companions | 'Thrive!' does well in the mixed border in the middle, or in a container where it is sure to shine. The true red of this rose is easy to match with many other colors. I recommend planting it with 'Brothers Grimm', 'Lady Elsie May', 'Postillion', 'Sweet Jane', and 'Tequila'.

Disease resistance	50
Flowering	27
Fragrance	0
Total rating	77

Disease resistance	49
Flowering	24
Fragrance	2
Total rating	75

Floribunda

Upright and bushy to 4–5 feet

Orange and coppery-pink

‘Thrive! Copper’ was previously sold as ‘Tequila Supreme’. Its present name accurately describes the exceptional color of its flowers, which is a unique blend of orange, pink, and a little bronze. As is typical of Floribundas, the flowers bloom in clusters and open to a semi-double form. The petals have a unique ruffled edge. This effect is nicely set off by the glossy foliage.

Introduced in 2009 by Meilland as ‘Jean Cocteau’, in 2011 as ‘Tequila Supreme’, and in 2014 as ‘Thrive! Copper’ (MEIkokan).

Companions | The unique coloring of ‘Thrive! Copper’ offers you the chance to create some fun color matches. Try it with pinks and other orange-peaches like ‘Belinda’s Dream’, ‘Cinderella’, ‘Eliza’, ‘Felicitas’, and ‘Michel Bras’.

'Thrive! Lavender'

Disease resistance 54
Flowering 28
Fragrance 3
Total rating 85

Shrub

Bushy and upright to 3 feet

Lavender

This award-winning rose was sold as 'Lavender Meidiland', but I think the name change to 'Thrive!' is apt, because that certainly describes what these plants do. The blooms seem to be in constant production, producing a lovely and pleasing lavender and pink whirl of color. The flowers are about 2 inches across and are produced in great numbers, sitting above the semi-glossy green foliage. I really love how these blossoms open up in small little cups to reveal their stamens. The finished blooms seem to fall cleanly and the whole effect is charming.

Introduced in 2009 by Meilland (MEIbivers).

Companions | With the lavender color and wonderful stamens, try planting 'Thrive! Lavender' with some orange-apricots such as 'Mandarin Ice', 'Michel Bras', 'Mother of Pearl', 'Sweet Fragrance', and 'Sweet Jane'.

'Thrive! Lemon'

Disease resistance	58
Flowering	30
Fragrance	0
Total rating	88

Shrub

Bushy, compact growth to about 3–4 feet

Light yellow

This prize-winning rose was sold as 'Limoncello' before being renamed as part of the Thrive! series. I can honestly say that this rose flourishes beyond anything a gardener could ask for. Here is another yellow that has superior disease resistance and just keeps on producing flowers all season. The blooms are very simple in their appearance, with five to eight petals surrounding a wonderful display of darker yellow gold stamens. As the flowers age, they fade to a much lighter yellow, even cream. The overall effect is a constant yellow and white patch of color in the garden.

Companions | 'Thrive! Lemon' has a compact habit, so place it near the front of the border. As the petals fade, the light color really complements the newer soft-yellow blooms, so try pairing it with some other light colors to really accentuate this effect. I recommend 'Easter Basket', 'Escimo', 'Innocencia Vigorosa', 'KOSMOS', and 'Old Baylor'.

Introduced in 2009 by Star Roses (MEljecycka).

'Topolina'

Disease resistance	56
Flowering	26
Fragrance	0
Total rating	82

Miniature

Short, spreading habit to 2 feet

Light pink with a white eye

When I first saw this little rose growing in the trial fields I stopped and stared, completely captivated by its simple charm. 'Topolina' means little mouse in Italian, and this is certainly a rose to put a smile on your face as it is so wonderfully cheerful and prolific in its bloom. The blooms are about 1 inch wide, with long yellow stamens perfectly presented against a lighter eye. This rose may be small for a shrub, but it won the ADR Rose Award in 2010.

Introduced in 2012 by Kordes-Newflora (KORpifleu).

Companions | 'Topolina' is great for mass plantings or spaces along pathways or in the front of the border. The color mixes well with a variety of other colors, especially those from a softer palette. The light eye shines beautifully when the plant is placed in front of cream and whites roses like 'KOSMOS' and 'Lion's Rose'. The yellow stamens will bring attention to other yellows such as 'Karl Ploberger', 'Solero', and 'Winter Sun'.

Floribunda

Bushy, spreading habit to 3–4 feet

Yellow

'Tupelo Honey' is a fun name for a fun rose. This variety has trialed very well in the heat and humidity of the South, proving its proven disease resistance in tough conditions. The blooms are a throwback to traditional roses, opening up to many-petaled sweetness. There is something about the color of this rose that is hard to pin down. Is it sunny yellow? Butter yellow? Honey yellow? Can it spread on a slice of bread! Delicious! Wherever you want to brighten up part of your garden, try this rose.

Introduced in 2014 by Kordes-Newflora (KORangober).

Companions | 'Tupelo Honey' has a habit that makes it well-suited to the middle to front of the border, and it shines in a container or in a mass planting. The yellow can be combined with other roses to create some hotter color combinations. Try it with 'Bajazzo', 'Garden Delight', 'Kew Gardens', 'Peach Drift', and 'Topolina'.

Disease resistance	44
Flowering	26
Fragrance	5
Total rating	75

'Wedding Bells'

Disease resistance	60
Flowering	28
Fragrance	6
Total rating	94

Hybrid Tea

Upright growth to 5–6 feet

Pink blend

As you can see by its high score, this is a fantastic and widely adaptable rose. I have been growing 'Wedding Bells' for a few years now and all I can say is "Wow." It is everything in a rose that I want. Even though it was only introduced a few years ago, 'Wedding Bells' has a certain maturity as a plant, by which I mean its genetic disease resistance, growth habit, and flower form are well and truly established.

The growth of 'Wedding Bells' is so surprising. It seems to break new growth from every nook and cranny and produces these gorgeous flowers. (This is a good example of a rose that you need to observe in order to determine how best to prune it.) The foliage is perfect—deep green and glossy. There has not been one speck of disease on this rose since I have had it in my garden. I am also amazed at the sturdiness and substance of the petals and the fullness of the bloom. They are the size of a baseball (or a cricket ball), fully rounded and multi-petaled. I think the bloom is most beautiful when it is not fully opened because it is almost sculptural in its form. Once open, it is a blend of variations on pink from the least to the most saturated, with a silvery sheen. Here is the icing on the (wedding) cake—everyone seems to enjoy the fragrance of this rose.

When describing these roses, I sometimes find it hard to come up with the words to give you the exact feeling I'm trying to convey. So to keep it short and simple, go out and plant 'Wedding Bells'.

Introduced in 2010 by Kordes (KORsteflali).

Companions | When color-matching roses in the garden, I aim for synergistic pairings. 'Wedding Bells' is a strong, upright grower so if I'm planting it with other roses, I put it in the middle of the border. It can be blended with other pinks, orange-peaches, blends, or creamy whites. Try it with 'Blush Noisette', 'Cinderella', 'Dee-Lish', 'KOSMOS', and 'Summer Memories'.

Large-flowered climber

Climbing to 5–8 feet

Red

Coming from Will Radler, the hybridizer of 'Knock Out', it is no surprise that this rose's attractive characteristics include disease resistance. I don't think I have ever seen any disease on it in my garden. The green, healthy foliage is a great backdrop for the red roses that always seem to be on this plant. Another winning characteristic is the fall color of the foliage and the plentiful rose hips that are a bright orange and persist to provide winter interest in the landscape. 'Winner's Circle' has a great blooming capability that drapes a fence, pillar, or arbor with continuous color, making a bold statement in your garden.

Introduced in 2008 by Conard-Pyle (RADwin).

Companions | Planted on a small fence, pillar, or arbor, 'Winner's Circle' provides a splash of color at any time throughout the year. Complement it with other landscape roses for constant, carefree rose success. I recommend 'Innocencia Vigorosa', 'Lady Elsie May', 'Oso Easy Cherry Pie', 'Sea Foam', and 'Sweet Fragrance'.

Disease resistance	**58**
Flowering	**23**
Fragrance	**0**
Total rating	**81**

'Winter Sun'

Disease resistance	52
Flowering	23
Fragrance	3
Total rating	78

Hybrid Tea

Compact, bushy growth to 3 feet

Light yellow

I'm happy to highlight some of the best, disease-resistant yellow roses in this directory, and 'Winter Sun' certainly fits the bill. The buds of this Hybrid Tea have the classic high-centered form, but they open into a more relaxed, cuplike shape that is very elegant in its own way. The flower color is a beautiful butter yellow that has lighter color on the outer petals. Overall, the leaves are green, healthy, and glossy. I have seen some leaf spot occasionally, particularly in poor weather. But this rose is tough and bounces back under its own strength. That's why it has received so many awards, including gold medals at the Le Roeulx Rose Trials (2010) and the Royal National Rose Society (2011), as well as First Prize at the Baden b. Wien–Vienna Rose Trials in 2011. For a yellow Hybrid Tea, I'll take it any day.

Introduced in 2010 by Kordes (KORbatam).

Companions | 'Winter Sun' has a more compact habit, so place it near the front of the border. The lighter outer petals really set off the still, soft yellow on the inside. Try to accentuate this effect by planting it with lighter-colored roses like 'Easter Basket', 'Escimo', 'Innocencia Vigorosa', 'KOSMOS', 'Macy's Pride', and 'Old Baylor'.

Disease resistance	55
Flowering	23
Fragrance	4
Total rating	82

Shrub

Medium, upright growth to about 4 feet

Lemon yellow

'Yellow Brick Road' is another strong yellow rose from the Easy Elegant collection. 'Yellow Brick Road' is one of my favorites in this series not only because I'm partial to yellow, but because of the really interesting bi-color effect of the lighter edges of the blooms. The lemon yellow color is difficult to convey in words, but it is so bright it lightens up a shadier area of my garden like a light bulb. With this bright lemon yellow comes a lighter edge to the petals that almost focuses the yellow color to the center, making it that much more intense. Set off by healthy, dark green foliage, this is a pleasing 'Yellow Brick Road' to follow, even in northern gardens.

Introduced in 2007 by Bailey Nurseries (BAload).

Companions | Place mid-sized 'Yellow Brick Road' in the middle to front of the border and combine it with 'Alister Stella Gray', 'Darlow's Enigma', 'Innocencia Vigorosa', 'La Perla', 'Macy's Pride', and 'Polar Express'.

'Yellow Submarine'

Shrub

Medium, upright growth to about 4 feet

Yellow

'Yellow Submarine' is also from the Easy Elegance series. It differs from 'Yellow Brick Road' in that the initial color isn't as bright (it is more buttery) and the flower tends to fade so that the petals show some light edges. The flowers are about 2 inches across, borne in clusters, and have a wonderful open, blousy effect. The disease-resistant, healthy green foliage allows this rose to sail in any garden.

Introduced in 2007 by Bailey Nurseries (BAline).

Companions | 'Yellow Submarine' is of medium height, so place it in the middle to front of the border. The coloration allows for some good combinations with other yellows and whites, so try it with 'Alister Stella Gray', 'Macy's Pride', 'Nastarana', 'Summer Memories', and 'Tequila'.

Disease resistance	54
Flowering	23
Fragrance	2
Total rating	79

Shrub

Taller, sometimes arching, and bushy to 5 feet

Pink

In bud form alone this is a gorgeous rose, and those lovely deep pink (almost wine-colored) buds open up to a very pleasant pink with dreamy and romantic flowers. The fully petaled and highly fragrant flowers even hold up well in the rain. The foliage is nice and green; although it has been susceptible to black spot at times, it usually bounces back after any infection. Even though it has won international awards, this is one of those sleeper varieties that are relatively unknown but deserve to be in more and more gardens.

Introduced in 2007 by Kordes (KORparofe).

Companions | 'Zaide' is a scented plant, so put it where you can relish the fragrance. There are so many tones of pinks in its blooms that almost anything is compatible with it, color-wise. Try a darker pink like 'Alexandra Princesse de Luxembourg', 'Dee-Lish', or 'Wedding Bells'. Lower-growing 'Larissa' and other whites like 'KOS-MOS' can be placed in front of it in the border or rose bed.

Disease resistance	54
Flowering	25
Fragrance	6
Total rating	85

Roses by class, habit, and color

'Kordes Moonlight'

'Awakening'

Roses by class and habit

China

'Ducher'

'Mutabilis'

'Pink Pet'

Climbing

'Above and Beyond'

'Awakening'

'Bajazzo'

'Brite Eyes'

'Climbing Pinkie'

'Florentina'

'Garden Sun'

'Golden Gate'

'Jasmina'

'Kordes Moonlight'

'Laguna'

'Morning Magic'

'Nahema'

'New Dawn'

'Peggy Martin Survivor'

'Rosanna'

'Winner's Circle'

Floribunda

'Black Forest Rose'
(upright)

'Brilliant Veranda'
(bushy, compact)

'Brothers Grimm'
(bushy, upright)

'Cream Veranda'
(bushy, spreading)

'Dolomiti'
(bushy, shorter)

'Easter Basket'
(bushy, shorter)

'Easy Does It'
(bushy)

'Elegant Fairy Tale'
(bushy, upright)

'First Crush'
(bushy, shorter)

'Fortuna Vigorosa'
(short, spreading)

'Garden Delight'
(bushy, tall)

'Innocencia Vigorosa'
(bushy, compact)

'Julia Child'
(bushy, shorter)

'KOSMOS'
(bushy, shorter)

'Larissa'
(bushy, spreading)

'Lion's Rose'
(bushy, upright)

'First Crush'

'Summer Romance'

'Mandarin Ice'
(bushy)

'Michel Bras'
(bushy, upright)

'Out of Rosenheim'
(bushy, upright)

'Petticoat'
(bushy, upright)

'Pink Bassino'
(bushy, shorter)

'Pink Martini'
(bushy, spreading)

'Plum Perfect'
(bushy)

'Pomponella'
(bushy, tall, and upright)

'Poseidon'
(bushy, upright)

'Raspberry Kiss'
(bushy)

'Ruby Ice'
(bushy, upright)

'Sister's Fairy Tale'
(lax, spreading)

'Solero'
(bushy, shorter)

'Summer Romance'
(bushy, upright)

'Summer Sun'
(bushy, upright)

'Super Hero'
(medium, upright)

'Sweet Jane'
(bushy, spreading)

'Sweet Vigorosa'
(bushy, shorter, spreading)

'Tequila'
(bushy, medium)

'Thrive! Copper'
(bushy)

'Tupelo Honey'
(bushy, spreading)

Grandiflora

'Centennial'
(bushy, tall)

'Mother of Pearl'
(tall, upright)

'Sweet Fragrance'
(smaller, upright)

Hybrid Kordesii

'Cape Diamond'
(medium, spreading)

'John Davis'
(climbing)

Hybrid Musk

'Darlow's Enigma'
(bushy, tall)

Hybrid Rugosa

'Thérèse Bugnet'
(tall, upright)

Hybrid Spinosissima

'Stanwell Perpetual'
(bushy, medium)

'La Perla'

Hybrid Tea

'Beverly'
(bushy, upright)

'Dark Desire'
(bushy, upright)

'Dee-lish'
(tall, upright)

'Eliza'
(tall, upright)

'Fiji'
(bushy, dense)

'Francis Meilland'
(tall, upright)

'Golden Fairy Tale'
(bushy, medium)

'Grande Amore'
(tall, upright)

'Heart Song'
(bushy, medium)

'La Perla'
(stout, upright)

'Marie-Luise
Marjan'
(bushy, upright)

'Savannah'
(bushy, upright)

'Souvenir de
Baden-Baden'
(medium, up right)

'Sunny Sky'
(bushy, upright)

'Thanksgiving Rose'
(bushy, upright)

'Wedding Bells'
(tall, upright)

'Winter Sun'
(bushy, compact)

Miniature

'Flirt 2011'
(compact, short)

'Lupo'
(bushy, dense)

'Oso Happy Petit
Pink'
(bushy, mounded)

'Roxy'
(bushy, small)

'Topolina'
(short, spreading)

Noisette

'Alister Stella Gray'
(arching, tall)

'Blush Noisette'
(bushy, tall, upright)

'Nastarana'
(bushy, tall)

'Rêve d'Or'
(climbing)

Polyantha

'La Marne'
(erect, medium, upright)

'Carefree Beauty'

'Quietness'

'Marie Daly'
(bushy, shorter)

'The Fairy'
(short, spreading)

Shrub

'Alexandra Princesse
de Luxembourg'
(arching, tall)

'All the Rage'
(bushy, medium)

'Belinda's Dream'
(bushy, upright)

'Caramella'
(bushy, spreading)

'Carefree Beauty'
(bushy, upright)

'Carefree
Celebration'
(bushy, upright)

'Carefree Delight'
(bushy, upright)

'Carefree Spirit'
(bushy, upright)

'Carefree Sunshine'
(bushy, medium)

'Carefree Wonder'
(bushy, upright)

'Cinderella'
(tall, upright)

'Coral Drift'
(short, spreading)

'Crimson Meidiland'
(bushy, spreading)

'Cubana'
(compact, spreading)

'David Rockefeller's
Golden Sparrow'
(tall, upright)

'Eifelzauber'
(tall, upright)

'Escimo'
(bushy, spreading)

'Felicitas'
(medium)

'F. J. Lindheimer'
(bushy, upright)

'Flamenco Rosita'
(bushy, medium)

'Flamingo
Kolorscape'
(bushy, shorter)

'Flower Carpet
Amber'
(bushy, shorter)

'Home Run'
(bushy, medium)

'Jane Bullock'
(bushy, compact)

'Kardinal
Kolorscape'
(bushy, shorter)

'Karl Ploberger'
(tall, upright)

'Kew Gardens'
(bushy, taller)

'Lady Elsie May'
(bushy, medium)

'Lady Pamela Carol'
(tall, upright)

'Lemon Fizz'
(bushy, rounded)

'Rosenstadt Freising'

'Thrive! Lemon'

'Lena'
(bushy, spreading)

'Macy's Pride'
(bushy, tall)

'Miracle on the Hudson'
(bushy, medium)

'My Girl'
(medium, upright)

'Old Baylor'
(tall, upright)

'Ole'
(bushy, spreading)

'Oso Easy Cherry Pie'
(bushy, spreading)

'Peach Drift'
(short, spreading)

'Polar Express'
(bushy, upright)

'Postillion'
(tall, upright)

'Purple Rain'
(bushy, dense)

'Quietness'
(tall, upright)

'Republic of Texas'
(bushy, compact)

'Roemer's Hip Happy'
(bushy, compact)

'Rosenstadt Freising'
(bushy, spreading)

'Sea Foam'
(bushy, rambling)

'Summer Memories'
(arching, spreading)

'Thrive!'
(bushy, medium)

'Thrive! Lavender'
(bushy, medium, upright)

'Thrive! Lemon'
(bushy, compact)

'Yellow Brick Road'
(medium, upright)

'Yellow Submarine'
(medium, upright)

'Zaide'
(arching, taller)

'Summer Sun'

'Yellow Brick Road'

Roses by color

Orange, apricot, peach, coral

'Above and Beyond'

'All the Rage'

'Caramella'

'Carefree Celebration'

'Coral Drift'

'Cubana'

'Easy Does It'

'Flower Carpet Amber'

'Garden Delight'

'Garden Sun Climber'

'Lady Elsie May'

'Michel Bras'

'Mother of Pearl'

'Peach Drift'

'Rosanna'

'Savannah'

'Summer Sun'

'Sweet Fragrance'

'Sweet Jane'

'Thanksgiving Rose'

'Thrive! Copper'

Yellow

'Carefree Sunshine'

'Centennial'

'David Rockefeller's Golden Sparrow'

'F. J. Lindheimer'

'Golden Fairy Tale'

'Golden Gate'

'Jane Bullock'

'Julia Child'

'Karl Ploberger'

'Kordes Moonlight'

'Lady Pamela Carol'

'Lemon Fizz'

'Postillion'

'Republic of Texas'

'Rêve d'Or'

'Solero'

'Sunny Sky'

'Tequila'

'Thrive! Lemon'

'Tupelo Honey'

'Winter Sun'

'Yellow Brick Road'

'Yellow Submarine'

'Alexandra Princesse de Luxembourg'

'Pink Drift'

Pink

'Alexandra Princesse de Luxembourg'

'Awakening'

'Bajazzo'

'Belinda's Dream'

'Beverly'

'Brite Eyes'

'Cape Diamond'

'Carefree Beauty'

'Carefree Delight'

'Carefree Wonder'

'Cinderella'

'Climbing Pinkie'

'Cream Veranda'

'Dee-lish'

'Dolomiti'

'Eifelzauber'

'Elegant Fairy Tale'

'Eliza'

'Felicitas'

'Fiji'

'First Crush'

'Flamingo Kolorscape'

'Flirt 2011'

'Fortuna Vigorosa'

'Francis Meilland'

'Jasmina'

'John Davis'

'Laguna'

'La Marne'

'Larissa'

'Lena'

'Marie Daly'

'Morning Magic'

'My Girl'

'Nahema'

'New Dawn'

'Oso Happy Petit Pink'

'Peggy Martin Survivor'

'Pink Bassino'

'Pink Drift'

'Pink Martini'

'Pink Pet'

'Pomponella'

'Quietness'

'Raspberry Kiss'

'Roemer's Hip Happy'

'Rosenstadt Freising'

'Roxy'

'Sister's Fairy Tale'

'Roxy'

'Zaide'

'Easter Basket'

'Macy's Pride'

'Souvenir de
Baden-Baden'

'Stanwell Perpetual'

'Summer Romance'

'Sweet Vigorosa'

'The Fairy'

'Thérèse Bugnet'

'Topolina'

'Wedding Bells'

'Zaide'

White and cream

'Alister Stella Gray'

'Blush Noisette'

'Darlow's Enigma'

'Ducher'

'Easter Basket'

'Escimo'

'Innocencia
Vigorosa'

'Kew Gardens'

'KOSMOS'

'Grande Amore'

'Brothers Grimm'

'La Perla'

'Lion's Rose'

'Macy's Pride'

'Marie-Luise Marjan'

'Nastarana'

'Old Baylor'

'Ole'

'Petticoat'

'Polar Express'

'Sea Foam'

'Spice'

'Summer Memories'

Red and Orange

'Black Forest Rose'

'Brilliant Veranda'

'Brothers Grimm'

'Carefree Spirit'

'Crimson Meidiland'

'Dark Desire'

'Florentina'

'Grande Amore'

'Heart Song'

'Home Run'

'Kardinal Kolorscape'

'Knock Out'

'Mandarin Ice'

'Miracle on the Hudson'

'Oso Easy Cherry Pie'

'Out of Rosenheim'

'Ruby Ice'

'Super Hero'

'Thrive!'

'Winner's Circle'

Purple and lavender

'Plum Perfect'

'Poseidon'

'Purple Rain'

'Thrive! Lavender'

'Poseidon'

'Thrive! Lavender'

Metric conversions

inches	cm
⅛	0.3
⅙	0.4
¼	0.6
⅓	0.8
½	1.3
¾	1.9
1	2.5
2	5.0
3	7.5
4	10
5	13
6	15
7	18
8	20
9	23
10	25
20	51
30	76

feet	m
1	0.3
2	0.6
3	0.9
4	1.2
5	1.5
6	1.8
7	2.1
8	2.4
9	2.7
10	3
20	6
30	9

Temperatures

$$°C = 5/9 \times (°F - 32)$$

$$°F = (9/5 \times °C) + 32$$

Resources

Suggestions for further reading

Bradley, Fern Marshall, Barbara W. Ellis, and Deborah L. Martin. 2010. *The Organic Gardener's Handbook of Natural Pest and Disease Control*. Emmaus, PA: Rodale Books.

Christopher, Thomas. 1989. *In Search of Lost Roses*. New York: Summit Books.

Lowenfels, Jeff. 2013. *Teaming with Nutrients: The Organic Gardener's Guide to Optimizing Plant Nutrition*. Portland, OR: Timber Press.

Lowenfels, Jeff, and Wayne Lewis. 2010. *Teaming with Microbes: The Organic Gardener's Guide to the Soil Food Web*. Portland, OR: Timber Press.

The Manhattan Rose Society. 2010. *The Sustainable Rose Garden: A Reader in Rose Culture*. Edited by Pat Shanley, Peter E. Kukielski,. and Gene Waering. Havertown, PA: Newbury Books.

Nauta, Phil. 2012. *Building Soils Naturally*. Austin, TX: Acres USA.

Olkowski, William, Helga Olkowski, and Sheila Daar. 1996. *The Gardener's Guide to Common-Sense Pest Control*. Coauthored and edited by Steven Ash. Newtown, CT: Taunton Press.

Rabhi, Pierre. 2006. *As in the Heart, So in the Earth: Reversing the Desertification of the Soul and the Soil*. Rochester, VT: Park Street Press.

Riotte, Louise. 1998. *Roses Love Garlic: Companion Planting and Other Secrets of Flowers*. North Adams, MA: Storey Publishing.

Scanniello, Stephen. 2005. *Rose Companions: Growing Annual, Perennials, Bulbs, Shrubs, and Vines with Roses*. Jackson & Perkins Gardening Guides. Minneapolis, MN: Cool Springs Press.

Walliser, Jessica. 2014. *Attracting Beneficial Bugs to Your Garden: A Natural Approach to Pest Control*. Portland, OR: Timber Press.

Zimmerman, Paul. 2013. *Everyday Roses*. Newtown, CT: Taunton Press.

Acknowledgments

I would like to thank the following family, friends, and colleagues for their help and expert knowledge, and for supporting my passion for disease-resistant roses.

Drew Hodges and the gang from SpotCo.

Marty Linsky and Lynn Staley, for your help and guidance.

To my Cushing friends for your love and support: Jody Payne and Jim Burgess, Anne and Mike Engelhart, Aura and Eli Ellis, Waite and Christine Maclin.

Bob Garst, I wouldn't be here if it weren't for you.

Thanks to the A.R.T.S team: Madeline Byrne, Mark Chamblee, Dr. Steve George, Gaye Hammond, Kerry Ann McLean, Daryl Otte, Michael Schwartz, Dr. David Zlesak, and Kathy Zuzek. Also thanks to all my great friends and colleagues on the Earth-Kind team.

To my friends in the rose industry: Christian Bedard, Michael Marriot, Alain and Matthias Meilland, Chris and Gary Pellett, Thomas Proll and Alexander Kordes, Will Radler, Mike Shoup, and Tyler Francis.

At Conard-Pyle, thanks to Jacques Ferare, Kyle McKean, and Martha Thompson.

To the gardeners and volunteers at the Peggy Rockefeller Rose Garden at the New York Botanical Garden, in particular Bernie Conway, Kenny Molinari, and Gene Sekulow and Jon Peter.

To my long-time rose friends: Mark Chamblee, Pat Henry, Sarah Owens, Gail and John Payne, and Pat Shanley.

Ginger Baxter, Barb Padezanin, and Scott Wheatley, thank you for your long-time support.

Thank you to Fiona Gilsenan for listening, for your guidance, and for putting up with me.

Many thanks to David Rockefeller for the wonderful opportunity.

Photo credits

Kordes Roses, pages 8, 47 left, 104, 119, 120, 134, 139, 170, 191, 196, 201, 209, 211, 218, 219, 224, 234, 235, 244 left, 245, 252 right
Matthias Meilland, page 144
New York Botanical Garden, pages 12, 14
Will Radler, page 237
Gene Sasse, pages 55, 105, 149
Star Roses and Plants/Conard-Pyle, page 109
David C. Zlesak, pages 67, 92, 187, 212

ALAMY
Mike Booth, page 75
Tabitha Fox, page 82

GAP PHOTOS
Christa Brand, page 43 lower left
Carole Drake, pages 37 bottom, 45
Manuela Goehner, page 43 top
Jerry Harpur, page 50
Lynn Keddie, page 42 right
Zara Napier, pages 20, 44 right
Friedrich Strauss, pages 42 left, 43 lower right, 44 left, 64
Graham Strong, page 48

SHUTTERSTOCK.COM
ArjaKos, page 57 upper right
Irina Borsuchenko, page 33
Colette3, page 47 upper right
Gerald A. DeBoer, page 78 lower left
BW Folsom, page 65
Ian Grainger, page 32
inlovepai, page 80 bottom
Karen Kaspar, page 73
Malgorzata Kistryn, page 57 lower right
Laura Mountainspring, page 78 upper right
periphoto, page 57 left
Bernhard Richter, page 78 lower right
Michael Solway, page 83

WIKIMEDIA
Arashiyama, page 30
Richard Avery, page 86
Scott Bauer (USDA), page 80 top
Cec-clp, page 76
SB Johnny, page 158
Gilles San Martin, page 85

All other photographs are by the author.

Index

About the Author

Ivo M. Vermeulen / NYBG

Peter E. Kukielski is widely recognized for his work toward sustainability and disease resistance in rose gardens. He was curator of the Peggy Rockefeller Rose Garden at the New York Botanical Garden from 2006 to 2013. In 2007, he redesigned the rose collection of their world-renowned Beatrix Farrand–designed garden. In 2008, he implemented a new mission: planting and trialing roses for disease resistance and less chemical usage in the garden. This book is a culmination of that research. Under Peter's leadership, the Peggy Rockefeller Rose Garden received the Great Rosarians of the World Rose Garden Hall of Fame Award for 2010 and was voted America's Best Public Rose Garden Display by the All American Rose Selections (AARS) committee. In 2012, the garden received the Award of Excellence from the World Federation of Rose Societies, recognizing it as one of the best rose gardens in the world. Peter came to New York from Atlanta, Georgia, where for more than ten years he owned and operated The Rose Petaler Inc., a rose garden design and maintenance business. Peter is currently on the National Earth-Kind team and leads the Northeast trials. He is also the executive director of the new American Rose Trials for Sustainability (A.R.T.S.), launched in 2014, national rose trials set up to scientifically determine the best roses based on regionality and climate. Peter is a contributor to and coeditor of *The Sustainable Rose Garden: A Reader in Rose Culture* (Newbury Books 2010). He resides in Portland, Maine. Further information about Peter can be found at millennialrosegarden.com.